# THE 50 BEST BIRDWATCHING SITES IN NEW ZEALAND

**Dedication**

This book is dedicated to Sam Tracy, my beloved of 33 years, who surprised us all by having a heart attack and leaving this world on 10 May, 2019. He was only 62. We visited so many of these birdwatching sites together. Sam driving, me taking notes and photographs, and both of us looking out for, and listening for, birds. We had long walks in the bush and the mountains, on beaches and over pastureland. We stopped for sandwiches and chocolate in stunningly beautiful places.

Sam is all through this book, it could not have happened without him, but he did not get to see it. He would have loved it. Forever and beyond, my love. And I am so grateful for those 33 years.

First published in the United Kingdom in 2019 by John Beaufoy Publishing Ltd
11 Blenheim Court, 316 Woodstock Road, Oxford OX2 7NS, England
www.johnbeaufoy.com



**Photo Credits**
**Front cover kererū** © Liz Light
**Back cover** Chatham Island Albatross © Mark Fraser/NZ Birds Online
**Spine** Rifleman © Oscar Thomas
**Title page** New Zealand Dabchick © Oscar Thomas



ISBN 978-1-912081-49-3

Edited by Krystyna Mayer
Designed by Gulmohur Press, New Delhi
Project management by Rosemary Wilkinson

Printed and bound in Malaysia by Times Offset (M) Sdn. Bhd.

# THE 50 BEST BIRDWATCHING SITES IN NEW ZEALAND

Liz Light

with Oscar Thomas Photography

JOHN BEAUFOY PUBLISHING

# CONTENTS

# Foreword

Each morning a tūī's song mingles with my dreams. Chime, cackle, shriek and trill, the tūī works through its ostentatious repertoire of both harmonious and discordant sounds. Its song is unique, unusual, and gives a lovely way to wake and join the day.

My office faces across a valley and, most days, a harrier hawk glides past, often at window height, smooth and unflappable, beady eyes looking for small moving things on the ground below. I keep a look-out for kererū/New Zealand Pigeons, big and pretty feathered, with iridescent wings, a glowing green breast and a white underbelly. In summer mating pairs fly high into the sky, then soar and swoop in great arcs, employing grand aerial acrobatics for the absolute pleasure of flying. Often a Pipit walks briskly on the lawn, tail constantly bobbing, looking for bugs and seeds and, recently, I watched a juvenile following a parent, learning the Pipit-feeding routine.

Before sundown I walk through the bush in the nearby regional park and am accompanied by Fantails, cheeky wee things, flibbertigibbeting around, swooping in, fantail spread wide, stalling mid-air to beak a small insect. They are tiny fearless birds, coming almost close enough to touch but too smart and fast for that. As Fantails flit around they squeak and tweet, a sweet, soft sound.

The night belongs to ruru, alias Moreporks, dark brown owls. At times they sit on the fence waiting for moths attracted to an outside light, and in a feathered flurry they grab their food and U turn in mid-flight back to the fence. They are night birds so it is not common to see them but every evening, before I sleep, I listen to their call. *Ruuuu ruuuu* echoes across the valley and fellow ruru perched in far forest respond. It is haunting; it is the sound of home, the sound of New Zealand.

Yes, there are skylarks that fill the sky with their fast tempo twittering in spring; chirping flocks of sparrows; an occasional thrush listening for worms on the lawn and a crazy blackbird who attacks his reflection on my window. But these are birds that have been introduced by settlers hankering for familiar homeland birds; they are in this remote southern land but not from it. It is New Zealand's endemic and indigenous birds that fascinate me and entice birding visitors to these southern islands. And some of these endemic birds are very special, totally unique species of the avian world.

Oscar Thomas

*New Zealand Fantail*

Gert op den Dries

*Ruru*

Liz Light

*Tūī*

# Introduction

New Zealand broke away from Australia about 80 million years ago and, in the way of plate tectonics, through time and infinitesimal increments, the separation between the two countries is now 1,700km. During that time there were many ice ages, massive volcanic explosions, major earthquakes and other catastrophic events, which caused the demise of land-living mammals that were here. These southern ocean islands belonged to birds and, without mammalian predators, birds evolved in unique and amazing ways. Many birds, such as the iconic kiwi and kākāpo, a nocturnal, flightless parrot, had no ground-dwelling predators to fly from so they lost the use of their wings and flourished as ground birds.

Eighty million years of separation gives birds plenty of time to evolve, and of the 91 species of terrestrial bird 85 are endemic – something that has occurred nowhere else in the world. Some of these, the beloved kiwi for instance, are absolutely freaky; their closest relative is the extinct giant Elephant Bird from Madagascar.

Liz Light

*Kākā*

Oscar Thomas

*New Zealand King Shag*

Then, merely 800 years ago, humans arrived in this place, Aotearoa, the long white cloud. Māori brought with them kiore, the Pacific rat, and kurī, small, wolf-like dogs. These mammals found flightless birds and their eggs easy pickings, and Māori, too, feasted on birds and their eggs.

Big, tasty, relatively easy to kill and flightless, the moa, all nine species of them, were extinct within 100 years of Māori settlement. With the demise of moa came the extinction of the great Haast Eagle, the biggest bird that ever flew; moa were its primary food source.

Oscar Thomas

*Rifleman*

Hunting, rats and dogs eliminated many smaller species of bird, too, and this, combined with habitat destruction, primarily by fire, did terrible work. By the time Europeans discovered these southern Pacific islands and renamed them New Zealand, 39 bird species were extinct.

Then along came European settlement with the introduction of the worst bird predators – more rat species, cats, stoats, weasels, ferrets, possums and different types of dog. And there was huge destruction of habitat by forest felling, farming, fire, wetland drainage, irrigation and river diversion in the settlers' unquenchable hunger for pasture and arable land. A further 19 species became extinct. The last bird extinction was probably the South Island kōkako, last seen in 1963, though hikers in remote mountain areas claim to have recently heard its exquisite chime-like call. Finding a few surviving pairs is not out of the question.

The New Zealand Storm Petrel was thought to be extinct for 100 years, when the sighting and photographing of one in 2007 caused a flutter in avian circles. In 2013 a nesting colony was discovered deep in the mountainous heart of predator-free Hauturu/Little Barrier Island. Likewise, the Campbell Island Teal was thought to be extinct until 1973, when a small population was discovered on a nearby islet. Predator removal from Campbell Island and a careful captive breeding and release programme have allowed this species to thrive.

Twenty-three bird species are considered critically endangered by the Department of Conservation (DOC). This means that they would not survive without human intervention and habitat management. Human intervention includes intensive predator trapping, some captive breeding and release to safe, predator-free environments, tagging, monitoring and a variety of other activities that will ensure the survival of a species.

Hundreds of thousands of New Zealanders are captivated by their native birds. Individuals, community groups, clubs and the green-leaning government are working together to rectify, or at least minimalize, the devastating consequences that land-use change, industrial farming, urbanization, irrigation and introduced predators have had on bird populations. It seems that the corner has truly been turned and there will be no more bird extinctions.

We Kiwis (New Zealanders are commonly referred to as Kiwis, a label we wear with pride) are taking up the challenge to support threatened species in ways that have been considered too hard a task in other countries and cultures. People realize that New Zealand's endemic flora and fauna is our greatest national treasure; it is part of our culture and part of what defines this distant, down-under land as unique, beautiful and different. The recent conservation efforts in New Zealand are an outstanding example to the rest of the world.

# Organization of the Book

*The 50 Best Birdwatching Sites in New Zealand* was written and researched by Liz Light but with knowledgeable input from numerous enthusiastic birders from different parts of New Zealand. The Auckland Ornithological Society has provided invaluable expert information, including that given enthusiastically by many terrific speakers at the monthly meetings.

The book is a well-organized love story devoted to our feathered friends which, when one really watches them, their habits, acrobatics and prettily patterned feathers, will brighten the dullest of days. We hope New Zealand's unique birdlife will not only intrigue armchair naturalists and travellers, but also encourage ecotourists and birdwatchers to visit New Zealand – and that it will help locals to enjoy their exceptional avifauna as they holiday in different parts of the country.

While the book is not intended as a photographic guide to New Zealand's birdlife, it includes around 345 images of birds and their habitats. We are extremely grateful to the devoted and talented individuals who have shared their photographs of the birds featured here.

The book is divided into sections: the introduction, true pelagic birds (tube-nosed seabirds) and penguins, and terrestrial birds and shorebirds at a glance, the systematic listing of land-based birdwatching sites, and the appendices.

The introduction contains a broad overview of New Zealand's geography, climate and bird habitats.

New Zealand has more pelagic seabird species than any other country – one third of the world's pelagic birds are found here, including many species of albatross. Pelagic birds are discussed in a separate section because, as seabirds, they spend much of their lives at sea and most of them nest on remote offshore islands. There are very few nesting sites on inhabited parts of New Zealand. Despite difficulties in accessibility you can see plenty of pelagic seabirds through specialist pelagic birdwatching cruises or day trips. Penguins are included in the pelagic section. They are pelagic, in that they spend most of their lives at sea, but they are not tube-nosed seabirds so are often not included in the category. Debate remains about this, but for the purposes of this book we include them in the pelagic section because birdwatchers are most likely to see them if they go on pelagic birding trips.

A brief overview of New Zealand's shore and terrestrial birds discusses the issues they face and what New Zealanders are doing to help them.

A separate section provides information on 50 accessible birdwatching sites. Each of the profiles includes such information as how to get there, the key bird species, tracks and trails, accessibility to the birds and the best time of the year to visit. Each includes profiles of selected birds. Note that the list of key bird species included with most site descriptions may not include *every* species that can be seen there.

Birds that are endemic, indigenous (that is, they breed in New Zealand but also breed in other countries) and self-introduced are included in the book. The 39 species deliberately introduced by humans, which have formed self-sustaining populations, are not included – these birds include sparrows, thrushes, blackbirds, quails, pheasants, starlings, skylarks and numerous ducks. Rare vagrants, birds that somehow got off-course and ended up here for a while but did not establish are also not included.

# Geography and Climate

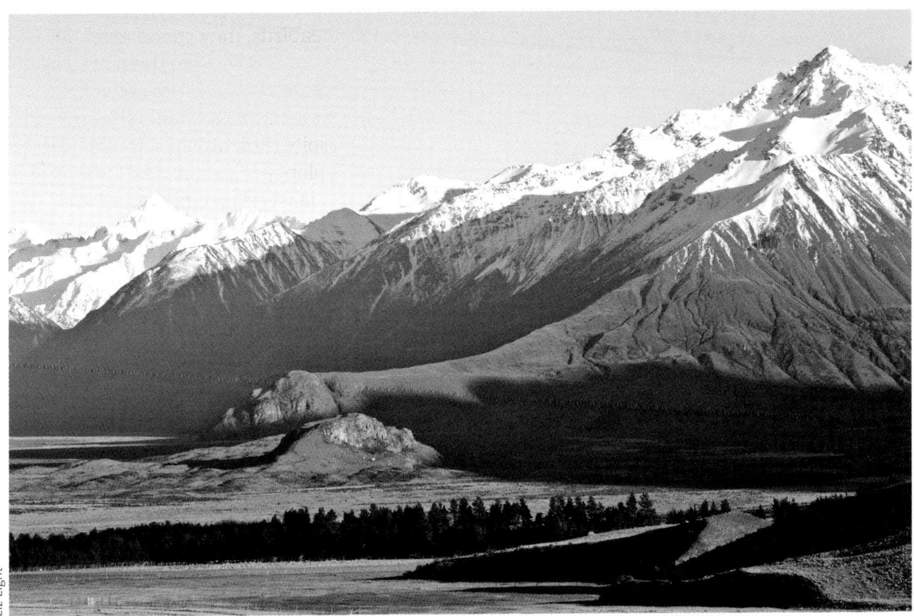
Liz Light

*The Southern Alps near Methven in the central South Island*

If a cosmic wizard picked up New Zealand and placed it in the same but opposite latitude and longitude in the northern hemisphere, it would stretch from Brittany in France to Casablanca in Morocco. If, in Auckland, you drilled a hole precisely through the middle of the Earth it would come out near Seville. This north–south length gives New Zealand a huge climatic range, from subtropical in the far north, to temperate verging on chilly in the deep south.

The geographic and climatic diversity is increased by New Zealand being situated on the edge of two great tectonic plates. The Pacific Plate and the Indo-Australian Plate, two massive slabs of the Earth's crust, meet beneath our feet. The meeting is not harmonious and the Earth, over thousands of years, rises, falls, buckles and tears. The downside of existing on the edge of tectonic plates is earthquakes and unpredictable volcanoes. The upside of living in these shaky isles is that the landscape is both varied and spectacular. The craggy, snow-covered alps have been pushed high into the sky by tectonic movements, and pointed, perfect-cone volcanoes have broken through thin, stretched parts of the Earth's crust.

There are also flat fertile plains, rolling green pastureland, deep, U-shaped glacial valleys, dripping rainforest, swampy brackish lakes and clear blue ones, harbours with mudflats and great tracts of sand dunes. This varied and diverse landscape provides different habitats for all sorts of bird species, but is conveniently parcelled in one long, thin country.

New Zealand has two main islands. It is 1,100km from the top to the bottom of the North Island and 900km from the top to the bottom of the South Island. Hundreds of other islands are also part of New Zealand. The land area is 268,000km$^2$ but the Exclusive

Economic Zone (EEZ), the greater territory belonging to New Zealand, including all distant islands and the sea between them, is 1.7 million square kilometres.

In the Southern Ocean, the subantarctic islands (Antipodes, Auckland, Snares, Bounty and Campbell Islands) are all World Heritage Sites and are enormously important for the world populations and diversity of seabirds (pelagic) (see page 17) and penguins (see page 25). Of the islands nearest to the mainland, and with kinder climates, 13 have long-term settled human populations. Rakiura/Stewart, Chatham, Great Barrier and Pitt Islands all have terrific birdwatching sites and are accessed by either scheduled air or boat services. Castle Rock, in the Dunstan Range, in the South Island, is the furthest place from the sea, being 120km from both the east and west coasts. In short, it is a long, thin country, the ocean is never far away and the influence of the ocean on the climate is ever present.

The predominant weather patterns move from west to east with alternating high and low pressure systems and the fronts between. The Southern Alps, in the South Island, stretch for 700km north-east to south-west. Aoraki/Mount Cook, in the middle of the Alps, is the highest mountain at 3,724m, and there are 16 other peaks over 3,000m in height. It is a substantial range of mountains and it cuts across the prevailing weather system. This means that there is heavy and regular rainfall in the west, and rain-shadow areas to the east of the Alps. The west-coast rainforest is spectacular, as are many of the birds that live in it.

The North Island's geography is more complex, with fault lines and massive areas of volcanism, both ancient and active, upsetting what might otherwise be clearer north-east to south-west mountain ranges. In the north the climate is influenced by subtropical storms that come down from equatorial regions, bringing with them thunderous rain and high winds.

In summary, the climate is temperate and moderated by the sea, and though there are four distinct seasons, the temperature seldom rises above 32° C in summer, or below minus five in the inhabited areas of the South Island in the winter. The weather is not sulky and rain does not linger for days. In some seasons brilliant sun, ferocious winds and heavy rain can all alternate numerous times in a day, as dark and dramatic clouds race in from the west.

Liz Light

*Lush farmland in East Kaipara*

Liz Light

*Lake Ellery near Jackson Bay in Westland*

# The Seabird Capital of the World

Oscar Thomas

*New Zealand Storm Petrel*

The two mainland islands of New Zealand are surrounded by hundreds of smaller islands dotted over a vast expanse of ocean. With the Kermadec Islands at 29 degrees south, Campbell Island at 52 degrees south and the Chatham Island group far to the east, New Zealand has a vast territory of ocean. In these waters seabirds forage, returning to their colonies, often on rocky, remote, windswept little islands, to nest each year. Forty-three per cent of the world's seabird species breed in New Zealand territory.

True seabirds, also called pelagic birds, are those that spend a significant portion of their lives in the open ocean and only come to shore to nest. They have long wings in proportion to their bodies and are powerful fliers. They can stay airborne for hours, effortlessly gliding and soaring, and many of them fly thousands of kilometres each year. They are biologically adapted to be able to drink sea water without salt building up to toxic levels in their bodies. This is done through a special salt-extraction gland on the top of their noses, hence the name tube-nosed seabirds. Seabirds usually nest in colonies, are monogamous and lay a single egg, and both parents are involved in the incubation and chick rearing.

The order Procellariiformes, tube-nosed seabirds, contains four families, 23 genera and 125 species. Of these, 53 species breed in New Zealand, and half of these are endemic. Of the world's 17 penguin (Spheniscidae family) species, six breed in New Zealand and a further five are regular visitors. With these statistics, New Zealand is absolutely the place to be for those interested in seabirds and penguins.

Oscar Thomas

*Southern Royal Albatross and White-capped Albatrosses*

# Threats to Seabirds

The New Zealand government, non-government organizations, and generous businesses and individuals have invested hundreds of millions of dollars and a huge amount of time in projects to eliminate introduced mammalian pests from the offshore islands. Doing so allows the ecology to return to something like it once was before humans arrived and brought bird predators with them, and gives birds safe, predator-free habitats in which to live and breed.

Since humans arrived, pelagic birds have had their nests destroyed by wild pigs and cattle, and their eggs and chicks eaten by cats, rats, hedgehogs and mustelids. Terrestrial birds have had their habitats, nests, eggs and chicks equally damaged by introduced predators.

Over time, the pest-elimination programmes have been extremely successful, with more than 100 of New Zealand's 220 larger offshore islands becoming predator and pest free. Consequently, many bird species are thriving and increasing in numbers, particularly terrestrial birds and pelagic birds that do not migrate to the northern hemisphere.

It is tragic, after so much effort being put into predator elimination, that many seabird species, especially those that migrate north during the austral winter, are returning from their annual migration to the north Pacific anaemic and weak, or are not returning at all. For instance, Flesh-footed Shearwater numbers are declining because of the diminishing food supply in the north Pacific.

The problems birds face in the north Pacific are many. Habitat for foraging shorebirds is being destroyed by mud-flat infill, pollution, industrialization and intensive net fishing. Global warming, and the consequent rise of the sea temperature, which some theorists say is higher in the northern hemisphere, is playing havoc with the food chain and changing birds' normal eating patterns, and industrial harvesting of krill is taking an essential element out of the food chain. Overfishing is also making it much harder for these high-revving, and consequently constantly hungry fliers to find the food they need. They spend more energy hunting for food and are debilitated by it.

The global fishing industry is another aspect of seabird loss. Albatrosses and many of the bigger petrels follow boats for fishy discards and get hooked by longlines or have wings broken by collision with trawl lines and boards.

Plastic is being mistaken for food by many seabirds. Ocean plastic pollution north of the Equator is worse than it is south of it because there are heavily populated countries in the north, many of which do not seriously bother with recycling or safe rubbish disposal. Many major rivers are gushing plastics and poisonous pollutants into the sea.

Then add nuclear radiation of the sea from the Fukushima tsunami and nuclear disaster. Some species that traditionally forage in that area are potentially affected by it. In summary, the future of migratory seabirds is not looking good.

These are big global issues beyond New Zealand's control, but I am proud that New Zealand has led the world in seabird bycatch mitigation techniques and practices in the fishing industry. It has also successfully demised introduced mammalian pests and predators in New Zealand's many Southern Ocean islands and a vast number of smaller mainland offshore islands. The colonies of pelagic seabirds, their home bases, are at least safe and secure.

# How to See Pelagic Birds

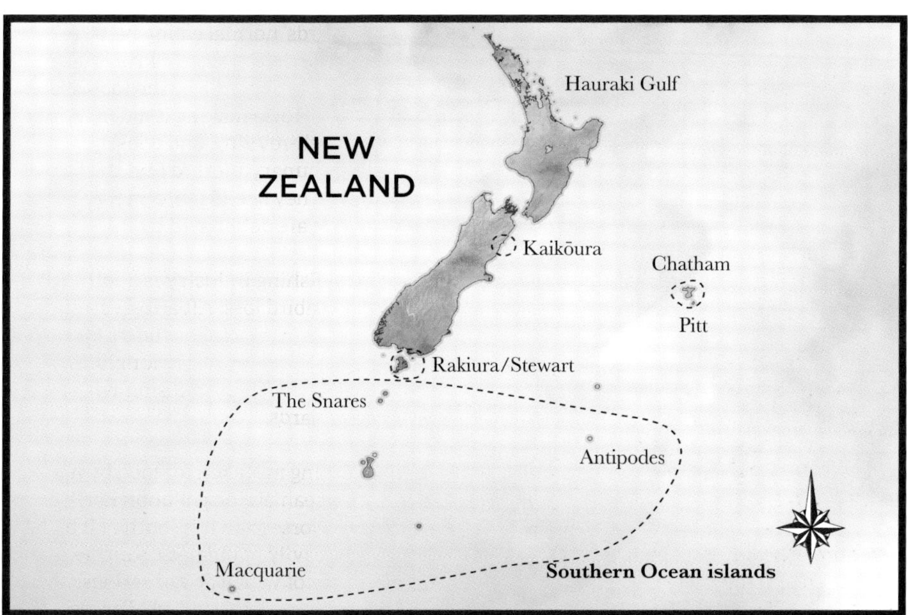


It can be challenging to see pelagic seabirds, and penguins, because most of them nest on remote offshore islands, and many of these islands are sanctuaries, so it is extremely difficult to get permits to visit them. The birds are also often flying far out at sea. However, there are numerous pelagic birding tour operators in different parts of New Zealand, with boat trips of various lengths from half a day to 30 days, for Southern Ocean cruises. The time of year is important as many of these birds are migratory and, keep in mind, seabirds have distinctly different ranges, habits and habitats. There are four accessible pelagic seabird hotspots.

The Hauraki Gulf, north-east of Auckland, has numerous islands, islets and rock towers, and the surrounding waters are prolific seabird nesting and feeding areas. Many of these islands are predator free and seabird colonies thrive on them. The area supports 27 species of seabird and includes the only known breeding colony of six species.

Hauturu/Little Barrier Island, 2,880ha, is a veritable treasure land for both terrestrial birds and seabirds. You cannot easily visit Hauturu and permits are required, but pelagic tour boats can get close. Likewise, the much smaller Mokohinau Islands support seven species of burrowing seabird. You cannot leave the boat but can get close to nesting sites for a marvellous pelagic birdwatching experience.

The Chatham Islands, an archipelago of 11 islands, of which two are inhabited, is a hotspot for both seabirds and terrestrial birds. There is one flight to the Chatham Islands each day from Auckland, Wellington or Christchurch (www.airchathams.co.nz). There are also small-plane flights to nearby Pitt Island, which has comfortable lodge accommodation. Mainland-based birding tour operators have scheduled visits to the Chatham Islands, but it is also possible to make your own way there, charter a boat for the day and get close to some of the birds on

Mark Fraser/NZ Birds Online

*Chatham Albatross and chick*

the smaller sanctuary islands (see page 211 for Chatham Islands).

Kaikōura, on the north-east coast of the South Island, is a pelagic seabird hotspot. An underwater canyon plunges a kilometre below the sea and this results in a vast seawater convergence zone and a rich food chain. It is a terrific habitat for whales, dolphins and seals, but there is also superb pelagic bird spotting.

Rakiura/Stewart Island, 37km south of the South Island, comprises one main island and numerous smaller ones. It is great for both terrestrial and pelagic birding. Many of the birds that breed on the Southern Ocean islands can be seen foraging here, as well as three species of penguin, in season. There is also very good terrestrial birding on nearby Ulva Island. There is a daily ferry to Rakiura from Bluff, a small-plane air service (www.stewartislandflights.co.nz) and a choice of accommodation. Numerous boat operators take birdwatchers on pelagic birding tours.

## Pelagic birdwatching tour operators

**Wrybill Birding Tours**, www.wrybill-tours.com. Wrybill Birding Tours has scheduled pelagic birding trips from four New Zealand ports and unscheduled ones from other destinations. Good for Hauraki Gulf and Chatham Islands.
**Heritage Exhibitions**, www.heritage-expeditions.com. Ship-based bird and wildlife expeditions to Fiordland, the Antarctic and subantarctic islands, numerous seasonal departures from 11 to 30 days.
**The New Zealand Seabird Trust**, www.nzseabirdtrust.com, organizes pelagic seabird boat trips in different destinations in the Hauraki Gulf numerous times a year. www.nzseabirdtrust.com/pelagic-trips.
**Manu Tours**, www.nzbirding.co.nz. Chatham Island birding tours as well as other destinations.
**Albatross Encounter Kaikōura**, www.albatrossencounter.co.nz. Scheduled half-day tours in this pelagic bird hotspot; full-day tours are available on enquiry.
**Rakiura Charters**, www.rakiuracharters.co.nz, does full-day and half-day pelagic birdwatching tours around Rakiura/Stewart Island.
**Aurora Charters**, www.auroracharters.co.nz, does full-day and half-day pelagic birdwatching tours around Rakiura/Stewart Island.
**Ruggedy Range Wilderness Adventure**, www.ruggedyrange.com, based on Rakiura/Stewart Island, for pelagic birds, penguins in season and more.
**Wilderness Lodge**, www.wildernesslodge.co.nz, Lake Moeraki, South Westland, for Fiordland Crested Penguins and more.
**Royal Albatross Centre**, www.albatross.org.nz, Taiaroa Head, Otago Peninsula for Royal Albatrosses and Little Blue Penguins.
**Elm Wildlife Tours**, www.elmwildlifetours.co.nz. Dunedin-based, for Yellow-eyed Penguins, albatrosses, seals and more.

# Modern Bird Families of New Zealand

## PELAGIC (TUBE-NOSED) SEABIRDS (ORDER PROCELLARIIFORMES)

### Albatrosses (Family Diomedeidae)

Of the world's 24 albatross species, 14 are indigenous to New Zealand and nine are endemic. Each species has different feaıres, habits and habitats and generalizations are difficult, but what they all have in common is that they are incredibly skilled at flying; they soar and glide seemingly effortlessly. Through satellite tracking it is known that albatrosses can cover 1,800km, point-to-point, in a day, and because they work the wind they actually cover much more distance.

### Great albatrosses (genus *Diomedea*)

There are four (possibly five) species of this king of birds. They include the Northern and Southern Royals and the Wandering Albatross. Wanderers include the Snowy Albatross, and Antipodean and Gibson's Albatrosses. There is debate as to whether the latter two wanderers are two separate species or two subspecies.

John Kyngdon

*Northern Royal Albatross*

Albatrosses have captured the imaginations of poets, artists and bird lovers. They are huge birds (up to 11 kg) and their 3m-plus wingspan, flying elegance, streamlined beauty and solitary independence in tempestuous oceans give them a mysterious fascination. Most breed on New Zealand's windswept Southern Ocean islands (Chatham, Antipodes, Auckland and Campbell Islands) and they circumnavigate the entire Southern Ocean, navigating with superb accuracy. They usually mate for life

Karyn Davis

*Southern Royal Albatross and chick*

John Kyngdon

*Gibson's Albatross*

and, after spending a year flying the world solo, they re-meet their sweethearts at the nesting site, and focus on chick rearing for 10 months. They can live to up to 80 years.

When the adolescents are courting and partners are reacquainting, they go through elaborate mating displays, throwing their heads back and shrieking, and slapping their beaks together, making frantic clapping sounds. Then, when the chicks are fledged, the parents leave and spend a year, alone, flying the Southern Ocean. If neither partner dies they return home and meet and nest again.

Both parents build the nest in spring, and only one white egg is laid, weighing up to half a kilo. The parents share incubation duty for 11 weeks until the chick hatches. Initially, one parent guards the chick while the other fishes. As the chick grows and requires bigger meals, both parents leave the nest to fish. In early spring, 10 months after mating began, the chick is fully fledged. It tests its wings and with a good breeze, glides off over the ocean – but it will return to its home base in five years to mate. Great albatrosses are not prolific; they take up to eight years to breed and a pair has only one chick every second year. If either parent dies during the 10-month baby-care period and the other cannot fish enough to feed the ravenous chick, it dies of starvation. In addition, terrible Southern Ocean storms wreak havoc on nesting sites, blowing eggs out of nests. Then add the destructive power of humans on fishing boats with longlines or nets, trawl boards and wires to the delicate balance. The result is that many albatross species are now vulnerable, some critically endangered. In fairness, the New Zealand fishing industry is super sensitive about this and mitigates against bird mortality by modifying its fishing techniques using numerous bird-scaring devices and fast-sinking hooks, and setting lines at night. Fishers from many other countries do not take such care, and albatrosses feed all over the Southern Ocean and in lower latitudes of the north Pacific.

John Kyngdon

*Salvin's Albatross*

Other threats include starvation of the adults and chicks through the ingestion of floating plastic. Another potential problem is the effect of climate change; already unusually hot summers have adversely effected the hatch rate of some species.

On the bright side, most of the nesting sites, New Zealand's Southern Ocean islands, are now, finally, pest and predator free. Eliminating introduced mice, rats, pigs, mustelids and cats from these remote islands has been a Herculean task. In Auckland Island, 46,000ha, the biggest and the last to be tackled, the elimination task is still under way. Antipodes, 11,000ha, has been declared predator free.

Pukekura Taiaroa Head is the only breeding colony of one of the great albatrosses (Northern Royal Albatross *Diomedea sanfordi*) on the mainland and it is easily accessible from Dunedin (see page 192).

**Mollymawks (genus *Thalassarche*)**

There are ten species of this albatross genus, seven of which nest in New Zealand.

John Kyngdon

*Buller's Albatross*

John Kyngdon

*White-capped Albatross*

The *Thalassarche* differ from the great albatrosses in many ways, mainly in that they are smaller, nest once a year and nest on pedestals of soil, guano and plant matter that can be up to a metre high. None nests on mainland New Zealand but they are often seen in summer, sitting on the sea, resting on a beach or following a fishing boat.

Mollymawks nest in the southern islands and disperse over the Southern Ocean in winter, some species circumnavigating it, some going to Chilean waters and others to the South Atlantic, and some staying in New Zealand waters.

Because these albatrosses nest once a year their populations tend to be less vulnerable than those of the great albatrosses (which nest every two years), though they suffer similarly with death and damage through the fishing industry. The Chatham Albatross *Thalassarche eremita*, for example, has only 4,500 breeding pairs, and with only one breeding site (The Pyramid, one massive rock near the Chatham Islands), bad storms at unsuitable times can be devastatingly damaging. The strength of storms is increasing with global warming.

Birdwatchers have a good chance of seeing Buller's, Salvin's, and White-capped Albatrosses on pelagic birding boat trips from Kaikōura and from Rakiura/Stewart Island, and Royal and Wanderering Albatrosses, depending on the season and luck on the day.

**Sooty Albatross (genus *Phoebtria*)**

This medium-sized (up to 3.7kg) albatross nests on Southern Ocean islands and forages further south than others of its species. It does not follow fishing vessels or scavenge as much as other albatrosses, and because of its shyness is seldom seen. Those who have had pleasure of seeing this handsome dark grey-all-over bird are impressed especially by its coordinated and acrobatic display while courting.

John Kyngdon

*Light-mantled Sooty Albatross*

## Fulmars, petrels, prions and shearwaters (Family Procellariidae)

Fulmarine Petrels, Gadfly Petrels, prions and shearwaters are all part of the family Procellariidae, in the order Procellariiformes (tube-nosed seabirds), which also includes albatrosses, storm petrels and diving petrels.

There are 72 species in the Procellariidae, of which 11 are endemic and a further 23 breed in New Zealand. The family is hugely diverse, with the giant petrels being as large as some albatrosses and other birds, such as prions, being tiny. Some are divers, others are surface feeders relying on small fish being pushed up by bigger fish, and some skim along the surface and filter out phytoplankton, krill and small crustaceans. The birds are colonial nesters, return to their home bases each year and tend to have long-term monogamous relationships. Some, the bigger petrels for instance, create scrappy nests, but almost all the smaller petrels, prions and shearwaters nest in burrows and, less often, crevices.

All species are long-distance foragers. Some of the Southern Ocean breeders move north to New Zealand mainland waters in the winter months, while most species migrate further north, over the Equator, to different parts of the Pacific. This book highlights those species that people on the mainland, and the accessible offshore islands, are likely to be able to see.

**Fulmarine petrels (subfamily Fulmarinae)**
There are only three species in this subfamily and the **Cape Petrel** *Daption capense* is common in New Zealand waters. Those lucky enough to see one will be impressed by the pretty black-and-white check patterns on its top sides. It nests in many subantarctic islands and ranges all over New Zealand waters in the winter months.

The **Northern Giant Petrel** *Macronectes halli* has a motley brownish, big, barrel body and is nearly as big as an albatross, but it lacks the long, elegant wings. It nests in and forages widely in New Zealand waters. The **Southern Giant Petrel** *M. giganteus* differs from its northern cousin only in the shape of its beak.

John Kyngdonw

*Cape Petrel*

John Kyngdon

*Northern Giant Petrel*

John Kyngdon

*Southern Giant Petrel*

**Gadfly petrels (genus *Pterodroma*)**
Eleven species of gadfly petrel breed in New Zealand. They are medium-sized birds with short bills and dark moulted plumage. They are noted for their distinctive flight, often described as speedy and weaving.

The **White-headed Petrel** *Pterodroma lessonii* breeds on the Southern Ocean islands but is often seen around the mainland in winter

Oscar Thomas

Cook's Petrel

Oscar Thomas

Broad-billed Prion

John Kyngdon

Black-winged Petrel

and spring. **Cook's Petrel** *P. cookii* breeds in two islands of the Hauraki Gulf, and since these islands have become predator free the populations are increasing. The birds forage around the mainland during the breeding season and migrate to the eastern Pacific Ocean after the chicks fledge.

**Pycroft's Petrels** *P. pycrofti* and **Black-winged Petrels** *P. nigripennis* are abundant and breed well on the predator-free offshore islands to the north and east but are difficult to see as the islands require visitor permits and the birds tend to forage in distant seas.

**Prions (genus *Pachyptila*)**

Four species of these small, blue-grey birds nest in New Zealand, in spring, in colonies and in burrows. They are monogamous and often mate for life. They nest on many predator-free islands around the coast and the population is in the millions. They can often be seen from the inter-island ferry, and from fishing and pleasure boats in coastal waters. Prions feed through surface diving or skimming across the water, consuming krill, small fish and crustaceans.

Oscar Thomas

*Black Petrel*

**Procellaria petrels (genus *Procellaria*)**
Four of the five species of these large, very dark petrels breed in New Zealand.

The **Westland Petrel** *Procellaria westlandica* is one of two petrel species that nest on the mainland. There is one nesting area in Paparoa on the west coast of the South Island. In the breeding season they can easily be seen by birdwatchers, at sunrise and sunset as they leave and return to their burrow nests. The population is stable due to intensive predator trapping in their nesting areas, but they are particularly susceptible to being killed during interaction with the fishing industry. They follow fishing boats and are superb divers, so even when fishers use mitigation techniques to rapidly sink baited hooks, these birds go after them. Night setting of hooks is necessary.

**Black Petrels** *P. parkinsoni* are big, gorgeous black birds that can be seen during the nesting season on Aotea/Great Barrier Island on the edge of the Hauraki Gulf. Like their Westland cousins, they are susceptible to becoming bycatch in the longline fishing industry, which is intensive in the outer Gulf where they forage. Their chicks are also susceptible to predation by feral cats and pigs, but this has declined in recent years through predator control.

**Shearwaters (genus *Puffinus*)**
This genus has 23 species, of which nine breed in New Zealand. They are medium sized and are so named because they skim across the water close to the surface. They have long wings often held at right angles to the body in a crucifix shape when flying.

**Buller's Shearwater** *Puffinus bulleri* nests in burrows on Poor Knights Islands to the north-east of Auckland, and is often seen in the Hauraki Gulf. During the breeding season it forages all over New Zealand's warmer waters and towards Australia. It has dark brown and grey patterned top sides and white undersides, and migrates to the north Pacific for the austral winter. It is a surface feeder on small fish and krill, so is unlikely to get hooked by longliners and seldom follows fishing boats.

**Flesh-footed Shearwaters** *P. carneipes* nest in burrows on 15 small islands around the North Island. They are a uniform dark colour and larger than Buller's and, unlike this species, they are diving birds and hang around fishing boats waiting for bait. Their numbers are

Oscar Thomas

*A boil-up of Buller's Shearwaters in Hauraki Gulf*

decreasing, not from land-based predation in their nest sites, but from interaction with the fishing industry, and plastic and pollution ingestion in eastern Japan where they migrate during the austral winter.

**Tītī/Sooty Shearwaters** *P. grisea*, also known as muttonbirds, are large and slender. They breed in burrows on hundreds of rocky and island locations. They are incredible long-distance fliers and cover vast arcs of the Pacific Ocean on the way to and from the north Pacific during the austral winter. Tracking reveals that some fly 74,000km a year and up to 500km in a day. They are also spectacular divers and can dive down to 60m for food, though they prefer surface feeding. They are known as tītī in New Zealand, and up to 250,000 fledgling chicks can be harvested by Māori (based on hereditary rights) each year from Rakiura/Stewart Island and the small surrounding islands. They are very fishy and oily and described as tasting like fishy mutton.

**Fluttering Shearwaters** *P. gavia* are small (up to 400g), brown and white, and have a distinctive flutter. They breed prolifically in northern New Zealand and are often seen in great rafts of sometimes thousands of birds

John Kyngdon

*Flesh-footed Shearwater*

Glenda Peake

*Sooty Shearwater*

*Fluttering Shearwater*

off the north-east of the North Island and in the Hauraki Gulf. They feed in association with school fish, and fish overharvesting could be detrimental to the population.

**Storm petrels (family Hydrobatidae)**
This is the smallest of the tube-nosed seabirds. Of the 20 species in the world, four breed in New Zealand and others forage in the country's waters. They are colonial nesters, in burrows or crevices, and feed on the surface (on plankton or small fish) by frantically fluttering just above the water. Most migrate to the northern Pacific for the austral winter.

The **White-faced Storm Petrel** *Pelagodroma marina* is common and breeds on many predator-free islands around the mainland and on the Chatham Islands. The small **New Zealand Storm Petrel** *Fregetta maorianus* was thought to be extinct until it was identified off Whitianga in 2003. There were subsequently more sightings in the outer Hauraki Gulf. In 2013 the nesting site was discovered in the centre of mountainous Hauturu/Little Barrier Island. The island has been predator free since 2004 and the population is secure and increasing. Fortunately storm petrels are much less likely to be killed in the fishing industry than other larger pelagic birds.

**Diving petrels (family Pelecanoididae)**
Of the five species in this family two breed in New Zealand. The **Common Diving-petrel** *Pelecanoides urinatrix* is abundant in coastal waters. This small, stocky bird is often seen in groups of a few hundred, and often flies close to the surface of the sea. It is particularly susceptible to predation by rats, cats and mustelids, but breeds prolifically on some 70 of New Zealand's predator-free offshore islands.

*White-faced Storm Petrel*

## PENGUINS (ORDER SPHENISCIFORMES, FAMILY SPHENISCIDAE)

Of the world's 17 penguin species, six breed in New Zealand, four only breed here and a further five are regular visitors. Three species nest on the mainland. Penguins do not fly and are not tube-nosed seabirds, but some consider them to be pelagic birds in that they spend most of their lives at sea. They belong to the family Spheniscidae, an endearing family that is taxonomically unique. In this book they are placed in the pelagic section as they face similar survival issues as other pelagic seabirds, and are almost as difficult for mainland birdwatchers to see.

Penguins are flightless, spending half their lives in the sea, to which they are brilliantly adapted. They are terrific swimmers, with flippers instead of wings, and blend with the ocean with their dark feathered top sides and pale undersides. They eat krill, fish and squid, which they catch while swimming. During the austral winter they live in the ocean, and recent tracking studies of 90 birds from two species show that they swim up to 15,000km in that time. Usually they return to their traditional breeding areas in spring to nest, head for the sea again when the chicks fledge, and return in early autumn for a few weeks to moult before going back to sea for half a year.

The **Little Penguin** *Eudyptula minor*, or kororā, the world's smallest penguin, nests in coastal regions of the mainland and on many Southern Ocean islands. It is easy to see, in season, at Tiritiri Matangi, Oamaru Blue Penguin Colony (see page 190) and Pukekura on the Otago Peninsula (see page 192).

The Fiordland Crested Penguin, Eastern Rock Hopper Penguin, Snares Crested Penguin and Erect-crested Penguin (subfamily Eudyptes) are all of medium height and have snazzy head markings, including a white stripe. In some of the species the eyebrow stripe can stick up ferociously when the birds are agitated.

The **Fiordland Crested Penguin** *Eudyptes pachyrhynchus*, or tawaki, standing some 60cm high, is the only mainland nester of this subfamily. It can be seen in South Westland, Fiordland and Rakiura/Stewart Island. The birds lay their eggs in July and August and continue to feed the chicks until early December, when they go to the sea, returning for three weeks in February to moult. They can be seen, in season, coming to shore in the early evening or leaving in the early morning at Monro Beach (see page 173), South Westland and on the north-west circuit hiking track on Rakiura/Stewart Island (see page 204).

Oscar Thomas

*Little Penguin*

Gerry McSweeney

*Fiordland Crested Penguin*

The **Yellow-eyed Penguin** *Megadyptes antipodes*, or hoiho, is the only species in its subfamily. It is a tall (75cm), handsome bird, with red feet, red bill, yellow eyes and a predominantly yellow head. It is also the most endangered of the penguin species, with estimates of only 2,000 pairs left. Around 500 pairs nest on mainland South Island, but the majority nest on Auckland, Campbell and Rakiura/Stewart Islands. Despite site management and predator control, mainland numbers continue to decline. There is concern that due to global warming and the increase in ocean temperature, the birds have to swim longer and harder to get food for their chicks. They can be seen, in season, on the Otago Peninsula (see page 194) and on Rakiura/ Stewart Island.

See also the list of boat operators for pelagic birdwatchers (page 16).

Liz Light

*Yellow-eyed Penguin*

## SHOREBIRDS, NON-PELAGIC SEABIRDS AND TERRESTRIAL BIRDS

### Bird habitats

Mainland New Zealand stretches from subtropical 34 degrees south to subantarctic 47 degrees south. Within these latitudes there are snowy alps, high up-thrust volcanic cones, pastureland, many thousands of kilometres of coastline, massive harbours with their attending mudflats, hundreds of windswept offshore islands and many different types of forest.

### Indigenous forest

In New Zealand there are 2,418 indigenous vascular plants, approximately 560 mosses, 2,300 species of lichen and 200 species of fern, from 10m-tall tree ferns to filmy ferns less than a centimetre long. Of these indigenous plants, 80 per cent of trees, ferns and flowering plants are endemic.

In the far north there are forests where tall kauri and kohekohe dominate. Mixed broadleaf and podocarp rainforest, in the central North Island, is dominated by other forest giants; rimu, miro, mātai, tōtara and kahikatea. In the crowns of the forest giants there are entire ecosystems of plant life – symbiotic, parasitic, creeping, ferns and more. This is a sky-high ecosystem for some forest birds. Underneath the forest canopy, especially in the wetter areas, there is a luxuriant growth of ferns, mosses, grasses and vines. The South Island has thousands of hectares of moss- and lichen-laden beech forest.

This rich and diverse habitat was an extraordinary place for New Zealand's unique birdlife to thrive in, but when humans arrived with fire, rats, dogs, cats and mustelids, paradisiacal habitats became dangerous; numerous bird species became extinct and others hover on the brink of extinction. Now only 15 per cent of the total land area is covered in indigenous flora, but that proportion is increasing as New Zealanders appreciate the rarity and value of their unique flora, its varied habitats and the importance of the indigenous flora to the birdlife. There are many predator-free islands and predator-fenced sanctuaries – mainland islands where birds live freely and thrive in large areas of indigenous forest.

Liz Light

*Broadleaf temperate forest, upper North Island*

Of the forest bird species, some are found in both main islands, and some live in specific parts of the North or the South Island. Subspecies of some of these birds have evolved in isolated areas. Large forest birds include weka, takahē, kererū, kākāpo, kākā and kōkako. Smaller forest birds include Riflemen, saddlebacks, Stitchbirds and honeyeaters, robins and Tomtits, New Zealand Fantails and Silvereyes, Whiteheads and Yellowheads, Brown Creepers, warblers, cuckoos and parakeets.

### Alpine environments

Some species are specifically adapted to living in the alpine environments, which include stumpy alpine shrubs, fields of herbs and flowers and high dry tussocks, especially in the South Island, including New Zealand Rock Wrens, kea, New Zealand Pipits, takahē and Great Spotted Kiwi.

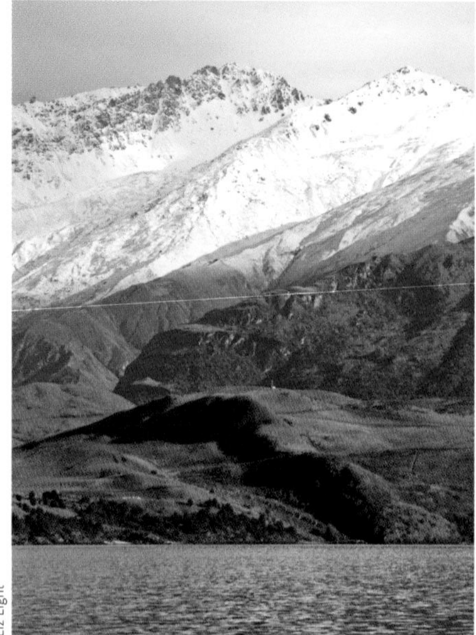
Liz Light

*Alpine habitats, Lake Wanaka, South Island*

Liz Light

*Estuaries and wetland, Waimea, near Nelson*

Liz Light

*Swamps and peatland, Lake Brunner, Westland*

Liz Light

*Braided rivers are found east of the Southern Alps*

### Wetlands

Approximately 30 per cent of New Zealand's indigenous birds live in or near wetlands. Sadly, wetlands have been radically reduced in the 170 years since European settlement. The hunger for farmland resulted in thousands of hectares of swamp, harbours and other wetlands being drained, reclaimed and otherwise damaged. In the last few decades the ecological value of wetlands has been recognized and the draining and polluting of them has ceased. Many wetland areas are being restored by wetland enthusiasts and birders. There are numerous types of wetland.

**Rivers and lakes** Due to abundant rainfall there are plenty of rivers and lakes. There are five indigenous duck species, three of which are abundant, and one, the Grey Duck, is in danger of being crossbred with the more aggressive introduced Australian Mallard. The rarest, the whio/Blue Duck, is one of the world's three species of torrent duck; it only lives in fast-flowing fresh water. Besides ducks, kingfishers, welcome swallows, pūkeko (swamphens), rails and crakes frequent lake and river edges.

**Swamps and peatland** Fresh and brackish water and the vegetation around it, in swamps and peatland, is the home of grebes, Black Swans, Paradise Shelducks, rails and crakes, pūkeko, kingfishers, fernbirds and kōtuku/White Herons.

**Braided rivers** Braided rivers are a unique geographical feature in the South Island. Mountain streams join rivers and, as they leave, fast flowing vertically off the mountains and arriving on the flat plains, they spread out in widely braided tendrils. Braided rivers are an inimitable and specific habitat for many finely adapted braided river birds. The Black-billed Gull, Wrybill, Black-fronted Tern, Banded Dotterel, kakī/Black Stilt, Pied Stilt and South Island Pied Oystercatcher are all river-shingle nesting birds. The Wrybill is a quirky wee bird, the only bird in the world with a laterally curved bill; it bends to the right, and is an ideal tool for prising insects out from under round stones.

**Mudflats, intertidal lagoons and sandbars**
There are vast harbours, hundreds of river mouths and intertidal lagoons providing numerous habitats, from mangrove forests (in the subtropical north), to fine sandy isthmuses, harbours and large areas of mudflats. The intertidal zones are important wintering grounds for thousands of Arctic nesting shorebirds. The most common are the Bar-tailed Godwit and Lesser Red Knot. Less common but still present each year are birds such as the Ruddy Turnstone, Eastern Curlew, Sharp-tailed Sandpiper, Terek Sandpiper, Red-necked Stint and Curlew Sandpiper. There are numerous species of endemic shorebird nesting only here, like the Banded Dotterel, Wrybill, Variable Oystercatcher, Pied Stilt, White-faced Heron and White-fronted Tern.

**Rocky coastlines and beaches** Harbours and river mouths are linked by extensive areas of rocky coastline, rocky islets and sandy beaches. This is the habitat of gulls, gannets, the New Zealand Dotterel and numerous species of tern.

Liz Light

*Beaches and dunes of the Kaipara Harbour, Northland*

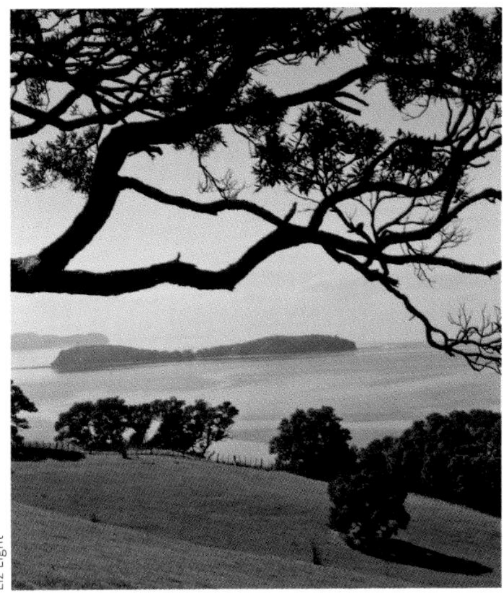
Liz Light

*Pastureland and offshore islands, Hauraki Gulf*

**Pastureland and urban areas**
European settlers brought many bird species to New Zealand, and about 34 species flourished and became part of New Zealand's avifauna. Introduced species are usually found in farmed pastureland, hedgerows and urban areas; habitat similar to where they came from in Europe. No introduced species thrive deep within the indigenous forest, though thrushes, blackbirds, hedge sparrows and chaffinches inhabit forest edges and compete with indigenous birds in that habitat. Skylarks, starlings, blackbirds and magpies are common pastureland birds. Common urban birds include sparrows, common pigeons, thrushes and blackbirds. Increasingly, endemic tūī and ruru/Morepork are found in urban areas as bird-aware people plant native trees in their gardens.

**Southern Ocean islands and other large predator-free islands**
As discussed in the pelagic seabirds and penguin section (see page 14) there are more than 100 large islands that are now introduced pest and predator free. This not only benefits the pelagic seabirds, but all sorts of terrestrial species are now thriving in these pristine environments. For instance, Antipodes Island has an endemic Antipodes Snipe, a pipit and two species of parakeet.

# Threats to Shorebirds, Non-pelagic Seabirds and Terrestrial Birds

Oscar Thomas

*Black Stilt*

### Water quality in wetlands, lakes and rivers

The water quality in many areas of swamp, peatland, rivers and lakes has been degraded by intensive dairy farming and dry livestock farming (beef and sheep). In forest and mountain areas the lakes and rivers are pristine, but it is downstream, when they pass through farmed pastureland, that the water-quality problems occur. The issues are specifically silt run-off, nitrogen and phosphorus from excess fertilizers and animal urine entering the streams, then rivers and lakes, and e-coli bacteria from livestock excrement. It is a dangerous mix that deadens water and creates a toxic environment for birdlife. The deterioration of water quality was a major election issue in 2017, and the end result was that the left-leaning Labour and Green parties won and numerous practices, such as fencing off waterways from stock and riparian planting along stream, river and lake edges to minimize fertilizer run-off are becoming compulsory. The intensification of dairy farming is being discouraged through the stopping of future irrigation projects.

### Climate change and global warming

Climate change is affecting bird habitats already but, as it amps up and rain patterns change, river and lake levels will increase or decrease and lake temperatures will rise. Already king tides regularly swamp the nesting areas of dotterels and other shorebirds. Combine a king tide with a storm surge during the nesting season, and the outcome is disastrous.

### Predator-free by 2050

The biggest threat to all shore and terrestrial birds is predation by introduced pests. Cats, dogs and mustelids (weasels, stoats and ferrets) kill adult birds when they can, and hedgehogs eat the eggs of ground-nesting birds. Flightless birds, such as kiwi, are particularly vulnerable to these predators, as are many other birds when sitting on nests at night. Eggs and chicks have little chance against these animals, and rats and possums also target ground nests. They are good climbers so nests in trees are not safe, either. These introduced mammalian predators have caused many species to become extinct. Others totter on the brink and would not survive in the wild without human management and

Oscar Thomas

*Fairy Tern*

*Takahē*

predator trapping. Fairy Tern *Sternula nereis* is critically endangered and doesn't respond to captive breeding. There are 40 to 50 birds left and only 10 breeding pairs. Massive effort is put into protecting the four nest sites but king tides and storms take them out as well as predators.

New Zealand leads the world in the successful eradication of predatory mammals from offshore islands, with more than 100 islands now being predator free and more being added to the total each year. The largest project so far, to rid the 50,000ha Auckland Islands of all introduced mammals (feral pigs, mice and cats) is under way but will take several more years.

Unlike islands, which are surrounded by sea that usually stops future infestation, making and keeping mainland areas predator free is a vastly greater challenge. It is a challenge that the people of New Zealand have taken on. Numerous predator-free mainland pockets have been created by surrounding specific areas with predator-proof fences, killing the predators within the area and preventing others from entering. Mountain Sanctuary Maungatauturi, 3,400ha surrounded by 47km of predator-proof fence, is the biggest mainland project so far. Besides this Zealandia Eco-sanctuary, Orokonui Eco-Sanctuary, Kaipupu Point Sounds Wildlife Sanctuary and both Tāwharanui and Shakespeare Regional Parks are predator free, with fences to keep them that way, and the increase in birdlife within these sanctuaries is astounding.

But this is not enough.

Predator-free 2050 is an ambitious goal to totally rid all New Zealand of the most damaging introduced predators that threaten the birdlife by eradicating rats, stoats, possums and wild cats. The project brings together central and local government, iwi (Māori tribes), philanthropists, non-government organizations, businesses, science and research organizations, communities, schools, landowners and individuals. To achieve this predators will be removed from bits of New Zealand one farm, one reserve, one suburb, one town, one city at a time – until those bits start to overlap and the predator-free map grows. The project has started, with Wellington City, Taranaki Regional Council and Waiheke Island being the first Local Bodies to commit to the process, and there are 34 other communities working on becoming predator free. This involves relentlessly walking and rebaiting trap lines, laying bait stations and monitoring predator numbers.

Having recently visited many of these predator-free bird sanctuaries, and seen and heard the density, variety and beauty of the endemic birds, I find the idea that all of New Zealand will, one day, be predator free hugely exciting. Predator-free 2050 is audacious, ambitious and possible.

*Black Robin*

# Terrestrial Bird, Shorebird and Non-pelagic Seabird Families

Eighty million years of separation from other land masses gave New Zealand's birds plenty of time to evolve, and of the 91 species of terrestrial bird in the country, 85 are endemic, that is, occurring nowhere else in the world. New Zealand also has the world's only alpine parrot, the kea, and the only flightless parrot, the kākāpo, which incidentally has the world's loudest call. The Wrybill, a small shorebird, is the only bird in the world with a bill that curves laterally; always to the right.

In the 19 orders of indigenous and self-introduced terrestrial birds, shorebirds and non-pelagic seabirds found in New Zealand (excluding vagrants), there are some 38 families, and within some of those families there are numerous species and subspecies.

Note that endemic, indigenous and self-introduced species are covered in this book, but not birds introduced by humans post-European settlement, such as thrushes, blackbirds, pheasants, quails, sparrows, common pigeons and more.

Key bird families are described below.

## Kiwi (Order Casuariiformes, Family Apterygidae)

Five species of this quirky and unique bird evolved in isolation over millions of years, and the kiwi is the much-loved icon of all New Zealanders.

This bird is so different from any others that avian systematizers still argue about its order and family. It is flightless and nocturnal, has whiskers and nostrils at the end of its very long bill, and its feathers are more like hair than feathers. Kiwi are monogamous and are believed to form lifelong bonds. The female lays one huge egg annually, the largest egg in the world in proportion to body size; some 20 per cent of it. Kiwi generally nest in burrows, holes in trees and other sheltered cavities. The chicks hatch fully formed, only stay in the nest for five days, then are out foraging with their parents, though they are not fed by them.

Kiwi can live for 50 years and grow to 25–45cm tall, depending on the species. Being nocturnal, kiwi are heard more than they are seen, and the male has a trilling, warbling cry with rising notes. The female is quieter and more gruff, and kiwi sometimes duet with the male taking the lead.

The **Northern Brown Kiwi** *Apteryx mantelli* is the most common species, though because of habitat loss, and predation by dogs, cats and stoats, the population, estimated to be up to 35,000, is declining in areas where there is no predator control. On some offshore islands and predator-controlled areas it is thriving. Tiritiri Matangi (see page 74) is one of many bird sanctuaries where this kiwi can be seen.

Oscar Thomas

*Brown Kiwi*

Sid Marsh

*Great Spotted Klwi*

The **Great Spotted Kiwi** *A. haastii* is the biggest of the species and stands up to 45cm tall. It lives in the north-west of the South Island, and some sources put the population at 16,000 and slowly increasing. It is often heard, if not seen, on the hiking tracks in the area.

Sallie Bassett

*Little Spotted Kiwi*

The **Little Spotted Kiwi** *A. owenii* was extinct on the mainland by 1980, but a small population on predator-free Kāpiti Island is thriving. Translocations to other offshore islands and predator-free inland sanctuaries have allowed the numbers of this little kiwi to increase. It can be seen by birdwatchers on Kāpiti Island (see page 141) or Zealandia Eco-sanctuary (see page 144).

**Tokoeka/South Island Brown Kiwi** *A. australis* inhabits three distinct areas of the lower South Island and Rakiura/ Stewart Island (see page 204), where it can easily be seen by birdwatchers. There are an estimated 25,000 birds in the wild.

**Rowi/Ōkārito Brown Kiwi** *A. rowi* is the rarest of the species with only some 375 remaining in a small area on the west coast of the South Island. Most birds live within the 11,000ha Ōkārito Kiwi Sanctuary and considerable effort is put into the conservation of these birds, including predator trapping and captive breeding and release. They can be seen, with the help of specialist birders, in the Ōkārito Kiwi Sanctuary (see page 168).

Chrissy Wickes

*Rowi*

## Swans and Ducks (Family Anatidae)

There are seven New Zealand species of duck and swan, and all are highly adapted waterbirds with webbed feet and flattened bills. They are mostly herbivorous, and some, like the Paradise Shelduck, survive well on pastureland.

The **Black Swan** *Cygnus atratus* is gorgeously elegant on water and when flying, but is a clumsy walker. It is a big bird, up to 140cm tall, and can have a 2m wingspan. It is found in groups in shallower lakes and rivers, including those in urban areas. Black Swans were abundant in New Zealand before human settlement but were hunted to extinction; the species was an easy food source for Māori. It later self-introduced from Australia and is now abundant, with flocks of thousands in some areas, for instance on Te Whanga Lagoon in the Chatham Islands.

Liz Light

*Black Swans*

The **Paradise Shelduck** *Tadorna variegata*, an endemic, is an attractive large duck that mates for life. The white-headed and copper-bodied female is, unusually for ducks, much more colourful than the plainer, dark male. Unlike other endemic ducks, the Paradise Shelduck has thrived post-European settlement and has adapted to living in pastureland and urban parks.

Oscar Thomas

*Paradise Shelduck (female in front, male behind)*

**Whio/Blue Duck** *Hymenolaimus malacorhynchos* is one of the world's three species of torrent-living ducks and it is rarely found far from a fast-flowing stream. It mates for life and the ducklings are fed by both parents. Numbers plummeted due to predation by introduced pests, especially stoats, and through competition for food with introduced trout. But the Whio Forever project, a combination of community volunteers, business sponsorship and DOC know-how, has stopped the decline. Numbers are increasing. This was done through intensive predator trapping and other forms of habitat management.

**Pāteke/Brown Teal** *Anas chlorotis*, an endemic, lives in lowland swamps and estuaries and enjoys forging on beaches near stream outlets. It is found in the wild in Northland in localized areas, where it is protected by intensive predator trapping. It is common on Aotea/Great Barrier Island, and thrives on some of the larger predator-free offshore islands.

Oscar Thomas

*Pāteke*

**Auckland Island Teal** and **Campbell Island Teal**, both small and flightless, are only found on the islands they are named after. Both faced extinction. Now that these large Southern Ocean islands have become predator free they are flourishing.

The **New Zealand Scaup** *Aythya novaeseelandiae*, a dear little duck, does well on big lakes and has benefited from the creation of the large hydroelectric lakes in both the North and South Islands. It is now quite common.

## Grebes (Family Podicipedidae)

New Zealand has two species of grebe. A local subspecies of the **Great Crested Grebe** *Podiceps cristatus* is found in New Zealand and Australia. It is a handsome bird with two parallel black crests on its head and chestnut frills on its cheeks. It is unusual (and cute) in that it carries its young on its back. It is only found in the large South Island lakes, and after declining in numbers for years, predator trapping and artificial nesting platforms have allowed the numbers to increase.

The **New Zealand Dabchick** *Poliocephalus rufopectus* is endemic, little and shy, so is not often seen. It lives on small lakes, ponds and sheltered inlets of larger lakes in the North Island. The population is declining as its nests are susceptible to boat wash and predation by introduced mammals.

Oscar Thomas

*New Zealand Dabchick*

## Gannets (Family Sulidae)

The **Australasian Gannet** *Morus serrator*, an indigenous species, is common around New Zealand's coast. It breeds in colonies on offshore islands and in isolated mainland areas. It is much admired because of its audacious diving. It plummets, straight as an arrow, into the sea. It has a pure white body, black wing-ends and a bright yellow head. It is easy to see, in season, at Muriwai gannet colony (see page 81) and Cape Kidnappers in Hawkes Bay (see page 134).

John Kyngdon

*Australasian Gannet*

**Shags (Family Phalacrocoracidae)**
New Zealand is a hotspot for shags. Of the world's 36 species, 12 are native to New Zealand and eight are endemic. They are commonly seen in coastal areas, though three species are also found inland, in freshwater environments. Shags live in colonies, usually nesting in trees and sometimes on dips in coastal rock, building untidy nests of twigs and seaweed cemented together with their faeces. Both parents build the nest, incubate the eggs and feed the young. They mostly eat fish, and are pursuit divers, but they also eat crustaceans and molluscs. Watching chick feeding is a great experience. A parent shag flies into the nest and two chicks, as big as her, flap, squawk and shove each other in an effort to get the fish. Mother chooses one and gives it the fish by opening her beak, with the chick putting its whole head down her throat in a bizarre snaking action. Following diving, it is a classic posture for a shag to stand on a rock and hold its wings out to dry, hence the common Kiwi saying, 'waiting like a shag on a rock'.

The **Little Black Shag** *Phalacrocorax sulcirostris* is common all over New Zealand. The adults have a white front and black top side, and stand up to 65cm tall.

The **Black Shag** *P. carbo*, also known as the Great Cormorant, is a big (up to 85cm), handsome black bird with an orange-and-white beak, cheeks and throat, and bright green eyes. It is common all over New Zealand, though it is sometimes found tangled with fishing lines and sinkers and sometimes drowns in set nets.

Oscar Thomas

*Pied Shag*

Oscar Thomas

*Little Black Shag*

The **Pied Shag** *P. varius*, also a large bird, can be found around much of the coastline. It is impressive looking, with blue eyes, a yellow patch on its head, primarily black top sides, and white undersides from the beak along the throat and belly to the underside of its tail.

The **Spotted Shag** *Stictocarbo punctatus* is undoubtedly the most glamorous in the family. It has two Mohican-type black crests on the top of its head, and bright blue-green skin that stretches from its blue-surrounded eyes to its long peachy beak. It has two wide, curved white stripes running from its eyes down either side of its neck and, during the breeding season, the grey plumage has clear black spots. It is abundant in most of New Zealand and breeds on rocky islands and mainland cliffs. It is a treat for bird photographers.

The **New Zealand King Shag** *Leucocarbo carunculatus* in Marlborough Sounds, and the shags of Rakiura/Stewart, Chatham, Bounty, Auckland and Campbell Islands, are all separate species but are rare and only found near the islands on which they breed.

Oscar Thomas

*Spotted Shag*

Oscar Thomas

*New Zealand King Shag*

## Egrets, Herons and Bitterns (Family Ardeidae)

In New Zealand there are six species in this family, besides vagrants. Herons and egrets have long, slender legs and a spear-like bill for fish catching, which they do by wading in shallow water, be it fresh, brackish or intertidal. They do not swim.

The **White-faced Heron** *Egretta novaehollandiae* is the most common of these birds. It is a tall bird with a blue-grey body and white face. It self-introduced from Australia in the 1940s, spread throughout New Zealand and is now abundant. It is most usually found in estuaries, swampy wetlands and even damp school playing fields after rain.

Glenda Peake

*White-faced Heron*

**Kōtuku/White Heron** *Ardea alba* has legendary status in New Zealand. It is a common bird in Asia but is rare in New Zealand, with only one nesting colony, near Ōkārito Lagoon, and a stable population of some 200 birds. After the chicks fledge they spread all over New Zealand, returning to the nesting site each August. Although this is the rarest of the herons in the country, it has a special place in the hearts of New Zealanders and is on the back of the $2 coin. It is loved for its elegance, beauty and size. It stands up to a metre tall, has bright white feathers and a yellow bill and, during nesting, grows a shower of long fine feathers on its back. It is a graceful flier with a 150cm wingspan, and flies with its head pulled back and a kink in its neck. Birdwatchers can view it from a shelter across a tidal creek (see page 168).

Liz Light

*Kōtuku*

The **Australasian Bittern** *Botaurus poiciloptilus* is found in large areas of swampland but is reclusive so is not easy to see. It is more likely to be heard than seen; the male makes a loud booming sound in the breeding season. The population is declining through habitat loss (draining of swampland), predation of chicks and eggs by introduced mammals, and decreasing water quality and reduced food availability.

Geoff Moon

*Australasian Bittern*

**Spoonbills (Family Threskiornithidae)**
There is only one species of spoonbill in New Zealand, the **Royal Spoonbill** *Platalea regia*. This splendid bird is a recent arrival, self-introduced from Australia, first noted in 1860 and now well established. There are 19 nesting colonies, including one among the kōtuku at Ōkārito. The birds feed in less than 40cm of water when the tide is right by swishing their spoon-shaped bills left and right.

Oscar Thomas

*Royal Spoonbill*

**Rails, Crakes and Swamp Hens (Family Rallidae)**
There are eight species in this family of swamp and marsh dwellers in New Zealand.

The **Banded Rail** *Gallirallus philippensis* is shy and well camouflaged, so is not easy to see. It frequents saltmarshes and mangroves in the upper North Island, Golden Bay and Rakiura/Stewart Island. It does well on the offshore predator-free islands, but on the mainland adults are preyed upon by cats, and the chicks and eggs by the other usual suspects. Banded Rails have red eyes, a pink bill and a dappled body with attractive stripes of orange, white and grey on the head and upper neck.

Glenda Peake

*Banded Rail and chick*

Oscar Thomas

*Spotless Crake*

The **Spotless Crake** *Porzana tabuensis*, a small rail, is a common but seldom seen bird in North Island freshwater wetlands and around thick reedbeds.

**Baillon's Crake** *P. pusilla affinis* is a small rail, half the size of a blackbird, which lives quietly and secretly in dense wetland vegetation. It is spread throughout New Zealand.

**Weka** *Gallirallus australis*, with four subspecies in different geographical regions, is an iconic flightless bird, appreciated because it is cheeky, curious and not afraid to approach people. Its call, when partners duet before dawn and after sunset, is easily identified and loved. It is a biggish bird (up to 1.4kg), dappled brown with pinkish legs and serious claws. The chicks and eggs suffer from predation, though an adult can see off stoats and cats with its powerful claws. The population is stable in areas where there is intensive trapping and on offshore islands, but otherwise is declining.

**Pūkeko** *Porphyrio melanotus* is a conspicuous bird, as big as a domestic hen, with long red legs, a red beak and purple-blue breast plumage. It is found everywhere there is fresh water and often in places where there is not, such as the grassy sides of the motorway north of Auckland. The chicks are cute: black fluff balls that strut around after their parents. Pūkeko are aggressive towards predators and other birds and are primarily vegetarian. They are admired and often depicted in cartoon form as a character in children's books.

**Takahē** *P. hochstetteri*, the world's largest rail (up to 2.7kg), looks like a big version of a pūkeko but has more spectacular back, breast and wing colouring. It is flightless and was thought to be extinct until rediscovered in a remote South Island mountain valley in 1948. It still lives in the wild in that valley, but captive breeding and release to numerous predator-free islands and mainland sanctuaries have made it more accessible for birdwatchers. There are now 350 of these treasured birds and the number is slowly increasing.

Gert op den Dries

*Weka*

Liz Light

*Takahē*

# WADERS, GULLS AND TERNS (ORDER CHARADRIIFORMES)

## Snipe-like Waders (Family Scolopacidae)

Bar-tailed Godwits, Red Knots and Ruddy Turnstones are New Zealand's most common trans-equatorial migratory waders. They nest in the north Pacific during the New Zealand winter, leave in March and return again in September. Though they are migratory, and do not nest in New Zealand, we love them and regard them as ours.

The **Bar-tailed Godwit** *Limosa lapponica* is in particular a bird that has national appeal. Each March godwits gather near tidal mudflats in the North Island before their long journey to Alaska, and groups of godwit fanciers also gather to say farewell to them and wish them a successful return in spring. Their departure and arrival each year is noted in the media; sometimes in the national news. They are large waders (up to 600g), with dappled brown feathers on top and creamy-grey below. They have long legs and long, slender bills.

Godwits are often seen at high tide, particularly before departure, in huge flocks in the northern harbours. They are extraordinary long-distance fliers and have been recorded as doing 11,000km without stopping in nine days. They make their way to tundra areas of western Alaska. When they return to their favourite New Zealand harbours in spring, they have been flying for so long that they are unable to fold their wings. They stroll around foraging with their wings limp, open by their sides, for a few days.

It is estimated that there are 75,000 godwits in New Zealand for the summer months but the numbers are dropping. The problem is not the harbour habitat in New Zealand, and they are not especially prone to predation. Godwits, like some other migratory waders, are seriously under habitat pressure in the feeding stop-off places on the East Asian-Australasian Flyway, especially around the Yellow Sea. Massive reclamation projects, pollution and overfishing are some of the problems. North Korea is the least developed country in the Yellow Sea and it is the only place where their Flyway habitat is still relatively safe and clean.

The **Red Knot** *Calidris canutus*, also known as the Lesser Knot, is not as big as the godwit and is sturdier, with a shorter bill and legs, but is often seen with godwits on shell banks at high tide. In February the birds moult to an attractive rusty-red on the breast and head, hence the name. About 30,000 spend the austral summers in New Zealand but the numbers are declining because of pressure on Bohai Bay, in the east of the Yellow Sea, where they stop to refuel on their way to their nesting grounds in Siberia.

Oscar Thomas

*Bar-tailed Godwit*

**Ruddy Turnstone** *Arenaria interpres* – to quote nzbirdsonline: 'This stocky blackbird-sized bird is a heavyweight-lifter among waders. Its stout body, short powerful legs and feet, and robust wedge-shaped bill allow it to turn over large shells, stones and flotsam, such as driftwood and seaweed, while foraging for sand-hoppers and other crustaceans.' Ruddy Turnstones are attractive, with white undersides and dark, intriguingly patterned top sides. Up to 3,000 come to New Zealand each austral summer, having nested in the Arctic tundra during the boreal summer, and they spread throughout the country mostly on estuaries and mudflats. As is the case with the other migratory wading birds, the biggest threat to Ruddy Turnstones is pollution and reclamation in the Yellow Sea, a staging stop on their journey to and from Siberia.

Oscar Thomas

*Ruddy Turnstone*

## PLOVER-LIKE WADERS (SUBORDER CHARADRII)

### Oystercatchers (Family Haematopodidae)

The **Variable Oystercatcher** *Haematopus unicolor* inhabits beaches, dunes and sandspits and is not migratory. The birds are monogamous and territorial, and often stay in the same areas, stridently defending their territories from people, dogs and other intruders, especially when they are breeding. They are glossy black, and have a long red bill and red eyes. The trapping of predators, which eat chicks and eggs, has benefited this species, whose numbers are steadily increasing with, now, an estimated population of 5,000 or more. They are common on beaches, especially on the east coast of the North Island.

Glenda Peake

*Variable Oystercatcher*

**South Island Pied Oystercatchers** *Haematopus finschi* look similar to their variable cousins but have a white underbody. Their habitat and distribution is totally different. They breed inland, on and near the beds of the South Island's braided rivers, and migrate to the far north in the winter. They are plentiful, with well over 100,000 birds.

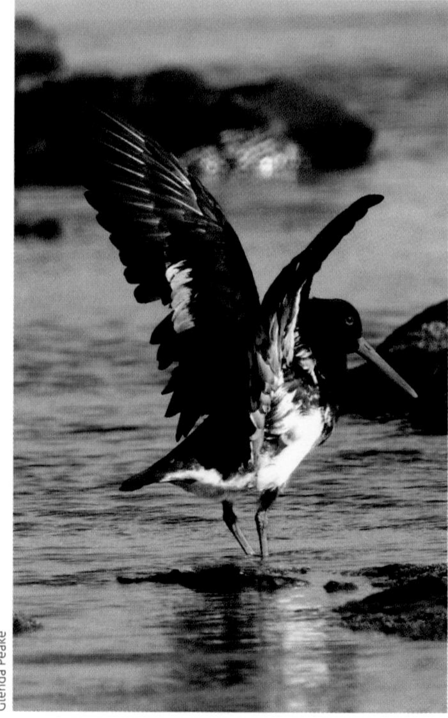
Glenda Peake

*South Island Pied Oystercatcher*

## Stilts (Family Recurvirostridae)

**Pied Stilts** *Himantopus leucocephalus* are black-and-white wading birds, long-legged and slender, which self-introduced from Australia in the early 1800s and thrived. There are now an estimated 30,000 living in wetlands including swamps, estuaries, rivers and lakes. They are aggressive towards predators, especially during nesting, hence the large and stable population.

*Pied Stilt*

**Kakī/Black Stilts** *H. novaezelandiae* have a black body, eyes and bill, and pink legs. Kakī are another back-from-the brink story. The species suffered terribly from predation and habitat loss until, in 1980, there were only 23 left in the wild. Three massive hydro lakes in the lower South Island, 60 years ago, cut into their habitat, and now they only breed in braided rivers, tussock grass and wetland of the Mackenzie Country. Due to a successful captive breeding programme there are now more than 100 birds in the wild, and other breeding pairs in aviaries. They suffer from predation by introduced mammals and, because they have a wide range, trapping and other forms of predator control are difficult. See faunarecovery.org.nz/kakiblack-stilt.

Oscar Thomas

*Black stilt*

## Plovers, Dotterels and Lapwings (Family Charadriidae)

The **Pacific Golden Plover** *Pluvialis fulva*, black-eyed, with dappled brown and golden top sides and creamy dappled undersides, is another of the migratory wading birds that breed in Alaska, and approximately 1,200 of a much greater population spend the austral summer foraging on tidal mudflats and estuaries around New Zealand.

**Shore Plovers** *Thinornis novaeseelandiae* are colourful little birds with orange-red beaks, a black face-mask, white breast and underparts, and grey-brown crown and upper bodies. They weigh about 60g and have relatively short legs for shorebirds. There are only 250 of these birds left, but before mammalian predators arrived they lived around most of New Zealand's shores. By 1960 there was only one population on a small predator-free island in the Chathams group. Captive breeding and release has halted the decline, but the birds remain susceptible to avian predation. The best chance of seeing them is on Motutapu Island near Auckland (see page 89), where they are successfully breeding each year.

Oscar Thomas

*Shore Plover*

The **New Zealand Dotterel** *Charadrius obscurus* is an endemic beach bird with, in breeding plumage, a peachy chest and spangled brown, cream and grey top sides. It lives on beaches in the northern North Island, especially on the east coast. The nest is in a scrape in the sand above the tide line and predators, especially cats and stoats, prey upon adults sleeping on their nests at night and on their tiny chicks. The nests are often swamped by king tides, which increase with global warming, and urban development along beaches infringes on their natural territory. The population is increasing in areas where predator control is managed but, otherwise, continues to decrease. Bring on Predator-free 2050 (see page 30), because there are only around 2,000 New Zealand Dotterels left and they are endearing little beach birds.

Oscar Thomas

*New Zealand Dotterel*

**Banded Dotterels** *C. bicinctus* are endemic and breed inland on riverbeds and gravel fans. They often migrate to estuaries after breeding. Banded Dotterels also struggle with predation, but there are around 50,000 of them. They are easiest for birdwatchers to see in estuaries or river mouths, at high tide during the summer, and are conspicuous because of the dark red band across their otherwise white breasts and a dark stripe on their throats.

Oscar Thomas

*Banded Dotterel*

The **Wrybill** *Anarhynchus frontalis* is a quirky endemic small plover. It is the only bird in the world whose beak turns laterally; it bends to the right. This makes the white-bellied, grey-backed Wrybill look a little odd, as if it bumped into something at speed and bent its beak. In fact, it is a clever adaptation to allow it to wedge under round river stones to extract insects hiding there. Wrybill breed in the braided rivers of Canterbury and, after breeding, from January to July, they migrate to the north primarily to the Firth of Thames and the Manukau Harbour. There are approximately 5,000 Wrybill, and it is thought that the population is slowly declining. The threats occur mostly during the nesting season and include predation, changing river systems through water depletion because of irrigation and weed growth, and recreational activities such as four-wheel-driving on the gravel beds. Birdwatchers can easily see Wrybill in northern harbours at high tide in summer.

The **Masked Lapwing** *Vanellus miles*, or Spur-winged Plover, self-introduced from Australia 80 years ago and is now all over New Zealand in open pastureland, riverbeds and urban parks. It has a harsh, shrill call and annoys horticulturalists by damaging leafy green vegetables. It is successful because it is aggressive. It vigorously, noisily and successfully defends its chicks against predators, and is well adapted to pastureland, of which there is plenty.

Oscar Thomas

*Wrybill*

## Skuas (Family Stercorariidae)

The **Southern Skua** *Stercorarius skua*, or Subantartic Skua, is a large brown gull-like bird. It breeds in the Southern Ocean islands and, in terms of habitat, has a lot in common with pelagic seabirds. It has a 1.5m wingspan and white-and-grey underwings, and weighs up to 2.2kg. Southern Skua disperse north during the winter and are sometimes seen in mainland waters.

The **Arctic Skua** *S. parasiticus* is a medium-sized gull (up to 600g) and a common migrant. It can be seen in coastal waters during summer, most often in the south and around the Chatham Islands. It often feeds by harassing other birds and forcing them to drop fish they have caught. It nests in Arctic tundra areas and heads south for the austral summer.

Oscar Thomas

*Arctic Skua*

Oscar Thomas

*Southern Skua*

## Gulls (Family Laridae)

**Karoro/Southern Black-backed Gull** *Larus dominicanus* is a large gull (up to 1kg). It has adapted to city living, nests on the tops of buildings and can be seen sitting on lamp-posts in central Auckland. It often scavenges by following fishing boats and hanging around beaches where people picnic, squawking for food. It is also common to see fully grown chicks begging from parents.

Liz Light

*Black-backed Gull*

The **Red-billed Gull** *Larus novaehollandiae*, an endemic, is a common coastal gull, often seen scavenging around fishing boats, rubbish bins and people with food on beaches and in cafe gardens. It usually breeds on cliffs and rocky stacks, and attacks those who pass close by. The population is declining in areas without predator control around nesting sites. Predators include cats, rats, stoats and ferrets, but predator control is increasing.

Gert op den Dries

*Red-billed Gull*

The **Black-billed Gull** *Chroicocephalus bulleri* is endemic and superficially looks like its red-billed cousin but is quite different. It primarily inhabits the gravel riverbeds of the South Island, seldom scavenges and is unlikely to be found in urban areas. The population is threatened from many angles. Introduced predators have easy access to the riverbed nests, irrigation has taken water from rivers so weed growth on gravel beds reduces the nesting sites, and drivers of off-road vehicles thoughtlessly crush the nests. The degradation of rivers and the environmental damage caused by intensive dairy farming were both recently huge election issues and the good guys won. This, along with a national awareness of mammalian predation of birds, will hopefully halt the slide in the population numbers of Black-billed Gulls.

Liz Light

*Black-billed Gull*

## Terns (Family Sternidae)

Of the 15 tern species found in New Zealand waters only two are endemic, one is a rare vagrant and 12 are indigenous. The Kermadec group of islands, 1,000km north-east of New Zealand, are all now predator free and are the home base of a number of tern species. The Brown Noddy, Black Noddy, Grey Ternlet, White Tern and Sooty Tern all breed on these islands. It is possible to visit the Kermadec Islands, but it requires a permit from DOC and a boat to get there.

Black-fronted Tern

Liz Light

Caspian Tern

**Black-fronted Terns** *Chlidonias albostriatus* are grey and small, and are seen around the braided rivers of the South Island and, when not breeding, at sea on the east coast below Hawkes Bay. They face the same challenges as other braided river nesting birds – predation, weed growth on gravel beds and flash floods. The population (above 5,000) is believed to be declining, but in recent years advocacy for terns and the problems they face has been increasing.

**Caspian Terns** *Hydroprogne caspia* are a native species found throughout New Zealand in lakes and harbours. They are common in the world but with only 1,500 breeding pairs they are not so common in New Zealand.

**White-fronted Terns** *Sterna striata* are endemic and breed around most of the coast on rock stacks, beaches, dunes, riverbeds and the piles of bridges and wharfs that are no longer used. They are medium sized, and have a long tail, and a black cap and bill. They often roost together at high tide on wharfs and moored boats. Most remain in New Zealand for the winter.

Liz Light

White-fronted Tern

## Hawks (Family Accipitridae)

There is one species from this family in New Zealand. **Swamp Harriers** *Circus approximans gouldi*, commonly known as hawks, are an abundant raptor, often seen at roadkill and perusing pastureland, gliding elegantly with slow beats of their long wings. They are big, with long legs and taloned feet, but these tuck under the tail when flying. They have a golden eye, hooked beak, greyish-brown feathers and long wings. The birds self-introduced from Australia in the 1860s and have benefited hugely from the destruction of forest and conversion of it to pasture. This species usually nests on the edges of swamps in reeds or rushes.

Oscar Thomas

*Swamp Harrier*

Gert op den Dries

*New Zealand Falcon*

## Falcons (Order Falconiformes)

The **kārearea/New Zealand Falcon** *Falco novaeseelandiae* has three subspecies in different geographies. It is New Zealand's fastest flier and can exceed 100km per hour when after prey. It usually lives in forest and mountain pasture and tussocks, and hunts live prey such as small birds, but it will go for rabbits, young hares and ducklings. It is now a protected species and the population is recovering. It is adaptable and has been introduced to vineyard areas, where it is useful as a natural way to keep small birds from eating ripening grapes. Falcons are not usually urban birds but a pair nests in Wellington and hunts all over the city.

## Pigeons (Family Columbidae)

The endemic **kererū/New Zealand Pigeon** *Hemiphaga novaeseelandiae*, and its relative, the **parea/Chatham Island Pigeon** *H. chathamensis*, are the only native New Zealand pigeons, though six other species have been introduced.

Jan Bottema

*Kererū*

Kererū are big (650g), magnificent birds whose dark green and brick-coloured back and wings have a glossy metallic shimmer. The belly is pristine white and the bill and eyes are red. The birds seldom vocalize, sometimes uttering a soft coo, but the swishing of their wings is distinctive. Kererū are birds that do terrific aerial acrobatics displays, seemingly at any time of the year, just for the fun of flying. Typically they fly very high, stall, swoop almost directly downwards, then sweep out of the stall in an impressive arch just before the ground. Kererū are common in most forest areas and increasingly in urban parks and gardens if there is an abundance of the fruit, flowers and leaves that they eat. Because they are large and eat whole fruit, they are an important vector in the spreading of seeds for the regrowth of native forest. They excrete seeds of up to 70 plant species and have been known to get drunk on over-ripe taraire berries and fall out of a tree.

### Parrots (Superfamily Strigopoidea) and Parakeets Superfamily Psittacoidea)

**Kākāpō** *Strigops habroptilus*, unique in the avian world, are peculiar but much-loved creatures. The arrival of humans and their mammalian predators almost caused the demise of the species and only 50 remained in 1980. In 2018 the population was 149. The population increased only after the birds were all transferred to three predator-free forested islands, and with intensive human intervention, especially during chick rearing, although they now also face a fatal fungal infection. Each kākāpō is named by the Kākāpō Recovery Programme, and though they roam freely each wears a radio transmitter to track its location. Kākāpō are very large (2kg) flightless night parrots with erratic breeding behaviour and may only breed once every four years, laying up to three eggs. Kākāpo are gorgeously green with mottled yellow and flecks of black.

Don Merton

*Kākāpō*

**Kākā** *Nestor meridionalis* are charming, rowdy and colourful parrots that live in both the North and South Island forests. They are medium-sized (450g), olive-brown parrots with glamorous red-orange underwings. They are often heard before they are seen, uttering a loud *ka kaa* cry as they fly high in the sky. They do well in forest areas where mammalian pests have been eliminated or are controlled. Kākā are common in suburban Wellington, having come there from the Zealandia Eco-sanctuary. They are mainly seed, fruit and nectar eaters, with excellent memories of the location of seasonal food sources. They are common on Aotea/Great Barrier Island, where they annoy the locals by raiding orchards. They are cheeky and sometimes silly, and are enchanting to watch.

Liz Light

*Kākā*

**Kea** *Nestor notabilis* lives only in the South Island and usually above 760m. It is the world's only true alpine parrot. Kea are closely related to kākā but are green not brown and have different vocalizations. They thrive in alpine forest and above the tree-line. They are intelligent, curious and not shy. Kea are famous for destroying the rubber on car-windscreen wipers in ski-resort car parks, will fossick inside an ignored backpack to steal sandwiches and cadge human food in other endearing ways. They are omnivores and will scavenge on animal carcases and prey on unfledged chicks of other birds, but also go for fruit, including berries, seeds and leaves.

Six species of parakeet are endemic to New Zealand and all of them are struggling to survive in the wild because of predation by rats, possums and stoats that climb trees and haul chicks and eggs from their nests, which are often in hollow trees. They thrive in the predator-free offshore islands and in specific areas of the mainland where there is intensive predator trapping. Unlike kākā and kea they are shy, live high in the canopy, and are difficult to see.

## Cuckoos (Family Cuculidae)

There are two species in New Zealand, the **Shining Cuckoo** *Chrysococcyx lucidus* and the **Long-tailed Cuckoo** *Urodynamis taitensis*. Both nest in New Zealand but migrate north for the winter. The distinctive whistle of the Shining Cuckoo is the avian announcement of spring's arrival. The Shining Cuckoo is found in most of New Zealand. The Long-tailed Cuckoo is rare and, sadly, becoming more so because of habitat loss in its Pacific island wintering grounds. Both species are nest parasites and lay one egg in the nest of other smaller birds. In the case of the Shining Cuckoo it is the tiny Grey Warbler that gets to hatch, then feed what becomes a ridiculously big chick in comparison to the warbler. Busy little warblers are adapted to this situation and they usually rear a clutch of their own chicks before the cuckoo lays its egg.

Oscar Thomas

*Shining Cuckoo*

## Owls (Family Strigidae)

There is only one species of native owl, **ruru/ Morepork** *Ninox novaeseelandiae*, and four subspecies of it. The owls are dark brown on top with chestnut and brown dapples on the underside. They have big yellow owl-eyes and a small hooked bill and, when roosting, have a rugby-ball shape. Some say that their two-note hoot sounds like *ruu ruu*, others that it is *more pork* – either way it is the sound of the night that means home to New Zealanders. They are blessedly abundant and there is nothing quite as delightful as parent ruru giving their young flying lessons, with three just-fledged chicks lined up nervously on a branch plucking up the courage to spread their wings and fly.

Gert op den Dries

*Ruru*

## Kingfishers (Family Halcyonidae)

The **Sacred Kingfisher** *Todiramphus sanctus* is the only species in this family that is native to New Zealand and it is abundant. It is often seen on power lines, high trees and fence tops above estuaries, ponds, rivers and streams, beady-eyeing whatever potentially tasty treat might move in the water. When it flies it is a fast-moving flash of buff and turquoise. It nests in burrows in clay banks, cliff faces and knots of trees. Kingfishers are strident defenders of their nests and will dive-bomb cats, rats and other birds that stray too near.

Oscar Thomas

*Sacred Kingfisher*

## SONGBIRDS (ORDER PASSERIFORMES)

In January 1770, while anchored in Queen Charlotte Sound, at the top of the South Island, Joseph Banks, learned naturalist on Captain James Cook's first expedition, wrote enthusiastically of being woken by the birds' dawn chorus: 'voices certainly the most melodious wild musick I have ever heard, almost imitating small bells but with the most tuneable silver sound imaginable'. Over the time of European settlement this rapturous dawn chorus all but disappeared as forest was felled for timber and then farming, and introduced mammalian predations wreaked havoc on a bird population not adapted to such predators. However, in the last 30 years, in many parts of New Zealand and the offshore islands, predators have been eliminated or are intensely trapped and the dawn chorus is returning. Beautiful and complex, it is rapturous to listen to and fun to identify the individual contributions of the avian songsters.

New Zealand has only 49 native passerines, 47 of which are endemic and nine of which are found only in the outer islands. There are numerous species of introduced passerine such as thrushes, blackbirds, starlings, skylarks, mynah and sparrows, but introduced birds are not covered in this book.

Oscar Thomas

*Rock Wren*

### Wrens (Family Acanthisittidae)

The Wren family, Acanthisittidae, is a very small family in New Zealand with just two tiny endemic members. The **Rifleman** *Acanthisitta chloris* weighs about 7g and is New Zealand's smallest bird. It is found in parts of the North Island and most of the forested areas of the South Island. The upper body is bright green on the male and yellowy-green on the female, and it has a tiny tail. It is monogamous and builds oval nests in cavities and hollows in trees.

The **Rock Wren** *Xenicus gilviventris* lives only in the alpine zone of the South Island. It is tiny, long legged and ground dwelling. Its feathers are olive-green on the top side and creamy below. Being a ground dweller it is particularly susceptible to predation but, thankfully, being alpine its nesting areas are beyond the range of rats, though mice and stoats prey upon its eggs and chicks.

### Wattlebirds (Family Callaeidae)

Not long ago there were four endemic species of wattlebird in New Zealand and now there are probably only three: one species has not been seen in over half a century.

The **North Island kōkako** *Callaeas wilsoni* is endemic and rare, living in 18 North Island locations. It is grey with bright blue wattles and a black beak and legs. It is a shy bird, referred to as the grey ghost but famous for its exquisite song. When kōkako duet the strong, pure sound stretches over many notes, like the chime of temple bells or the ring of a piano tuner's fork. Sadly the **South Island kōkako** is thought to be extinct, with

Oscar Thomas

*Kōkako*

the last confirmed sighting in 1963. Hikers in the remote mountains behind Nelson swear they have heard its beautiful call and others say they have sighted it, but it has not been successfully photographed. It has/had orange wattles but otherwise looked like its North Island relative.

Liz Light

*North Island Saddleback*

The **North Island Saddleback** *Philesturnus rufusater* and **South Island Saddleback** *P. carunculatus* are both endemic, and both threatened, but with predator trapping and management, the populations are making a recovery. These saddlebacks are the size of a blackbird, and have a dark body with a dark orange saddle over the back and similarly coloured wattles.

## Stitchbirds (Family Notiomystidae)

**Hīhī** *Notiomystis cincta* is endemic and unique, and is the only New Zealand member of the stitchbird family. This bird was prone to predation and suffered more than most birds from avian diseases and parasites that arrived in New Zealand along with introduced bird species. By 1980 there were no surviving hīhī on the mainland and one remaining population on Hauturu/Little Barrier Island. Due to good conservation management, research and appropriate translocations, hīhī are thriving in seven predator-free bird sanctuaries in the North Island. Hīhī are a conservation success story.

## Gerygones (Family Acanthizidae)

There are two closely related species in this family.

The **Grey Warbler** *Gerygone igata* is barely bigger than New Zealand's smallest bird, the Rifleman, miraculously thrives in a challenging environment (habitat loss and predation) and is common throughout New Zealand. It lives in forest and forest edges, and regrowth mānuka is a favourite habitat. It is grey-green above and off-white below, nothing special to look at but its song is a loud, sweet trill. It hosts the parasitic Shining Cuckoo, which lays its eggs in the nests, and the big cuckoo chick throws out the tiny warbler chicks so it alone is fed by the busy warbler parents. It is likely that warblers get a clutch away before the arrival of the cuckoos in spring, hence keeping their own species well populated.

The **Chatham Islands Warbler** *G. albofrontata* is a tiny hero. Conservators successfully used it as surrogate mother to the critically endangered Black Robin in the 1980s, when only one fertile female Black Robin survived. The Chatham Islands Warbler is endemic to the Chatham Islands, but struggles with harsh winters which knock back the population.

Oscar Thomas

*Grey Warbler*

Bellbird

## Honeyeaters (Family Meliphagidae)

The world has 182 species of honeyeater and New Zealand just two. They are both endemic and are both extraordinary songsters, the cornerstones of the choir that makes New Zealand's dawn chorus.

The **Bellbird** *Anthornis melanura*, alias korimako, is endemic to both islands and is widely spread in some localities. Populations from predator-free islands have recolonized areas of forest on the mainland now protected from predation. This is a small bird with olive-green colour variations and black on the tail and wings. It is primarily a nectar eater, but also enjoys fruit and insects. Its song, usually that of the male, is superb; the bell-like notes are clear and pure, and it has great pitch and tonal variation for such a small bird. The birds often gather in groups of three or more, fluff themselves up and sing in a heavenly chorus.

Liz Light

*Tūī*

**Tūī** *Prosthemadera novaeseelandiae*, found in all of the North Island and much of the South Island, is another iconic songbird. It combines bell-like notes with rasps, creaks and thumps, and in many areas this is the sound that wakes people in the morning. It is dark – black, navy blue and brown – with a petrol-on-a-puddle iridescent sheen. Tūī have a large tuft of white feathers on their throat, hence the name parson bird. They are aggressive defenders of their territory and their nests are big enough (30cm long) to mean business. They are nectar feeders by preference but go for fruit in the autumn. They are returning to suburban areas as New Zealanders plant nectar-producing trees in their gardens and offer sugar-syrup feeders.

## Creepers (Family Mohouidae)

There only three New Zealand species in the Mohouidae family, and they are all endemic small birds.

The **Whitehead** *Mohoua albicilla* and its close relative, the **Yellowhead** *M. ochrocephala*, or mohua, do not have overlapping ranges. Their appearance differs in that one has a white head and white underparts and the other yellow. The species thrive in predator-free areas, and translocation to islands and predator-free mainland sanctuaries has proved to be successful. The birds are vocal and fly about in groups tweeting, chirping and trilling.

**Pīpipi/Brown Creepers** *M. novaeseelandiae* are small forest birds found only in the South Island in a variety of forest types and scrubland areas. This is a squeaky tweeter and often flies in chattering flocks, providing a staccato addition to the dawn chorus. The birds build compact, deeply cupped nests of twigs and moss lined with grass, feathers and wool, and often have two clutches in spring and early summer.

Oscar Thomas

*Brown Creeper*

## Fantails (Family Rhipiduridae)

### Pīwakawaka/New Zealand Fantail

*Rhipidura fuliginosa* is endemic, much loved and abundant. Tiny, with a proportionately very large fantail, the birds are not shy of people and will accompany them on walks, flitting hither and thither in seemingly erratic ways while softly, incessantly tweeting. Despite predation they have fared well since European settlement as they can adapt to native forest, introduced commercial pine forest, scrubland, hedges and even suburban gardens. They do not do well in severely cold or wet winters. Their aerial acrobatics are fascinating to watch; pīwakawaka are smile makers.

Glenda Peake

*New Zealand Fantail*

## Australasian Robins (Family Petroicidae)

This is a small, sweet family of four endemic species.

The **New Zealand Tomtit** *Petroica macrocephala* has five subspecies on different islands. This wee charmer weighs just 11g, has a large head relative to its body and is prettily black and cream. It has a tendency to perch at an angle on tree bark and has a warbling whistle. These hard-working little birds often produce three broods a year, and this seems to keep their population stable despite predation. Like

Oscar Thomas

*New Zealand Tomtit*

*South Island Robin*

most birds, they do well in predator-free areas of forest and scrubland. They are not shy and you can get within a metre or two of them on forest tracks.

The **North Island Robin** *P. longipes* and **South Island Robin** *P. australis*, two similar subspecies, have sooty upperparts, a creamy breast and long, slender legs. They are small birds but bigger than Tomtits and warblers. They are not shy of people and will forage on the bush tracks at the feet of walkers. Their song is a many-noted cheep with an occasional trill.

The **Black Robin** *P. traversi* is a conservation success story. These habitat-particular birds, endemic to five of the Chatham Islands, were, post-European settlement, rapidly demised by predation and habitat loss. By 1980 their population had plummeted to just five birds and only one female. Conservation officers moved them to a predator-free environment with suitable forest and began a careful breeding programme. The heroine, Old Blue, laid two clutches of eggs a year. The first clutch was removed and raised by a surrogate mother, a Tomtit, so Old Blue laid again and raised the second clutch herself. She did this for a number of years and saved the species, which now numbers around 300 and lives on two predator-free sanctuary islands. In time, when other predator-free habitats are secured, Black Robins will be reintroduced to other Chatham locations.

## Fernbird (Family Megaluridae)

The **Fernbird** *Bowdleria punctata* is the only species in this family in New Zealand. It is endemic and widely spread throughout wetlands including reedbeds, saltmarshes, swamps and damp scrub or damp forest. Numbers have decreased because of the draining of wetlands for pasture and because of predation by introduced mammals. The birds are poor flyers and prefer to scramble. They do not exactly sing but utter a raspy, repeated tick with, sometimes, an intermittent squeak. They are medium sized (24g), and have creamy underparts and chestnut upper bodies with chocolate dark spots and semi-stripes. They are attractive and shy and blend well into their habitat.

Oscar Thomas

*Fernbird*

New Zealand Pipit

## Pipit (Family Motacillidae)

The **New Zealand Pipit** *Anthus novaeseelandiae* is endemic and there are three subspecies on distant offshore islands. Slender, brown and medium sized (35g), this songbird is easily identified because of the constant flicking of its tail when it forages. It is a ground forager and is bigger and more upright standing than sparrows, but has a similar colouring. Pipits enjoy open habitats such as scrub, poor pasture, coastlines and alpine shrubland.

## White-eyes (Family Zosteropidae)

The **Silvereye** *Zosterops lateralis*, also known as a wax-eye or tauhou, is a small bird (12–14g) and a relatively new arrival. It self-introduced, presumably from Australia, in the 1850s, probably blown to New Zealand on the strong predominant west winds. It is now common. Silvereyes have mainly olive-green plumage above and are cream below, with pale yellow heads, and are easy to identify because of the white ring around their eyes. They have a busy little twitter when foraging together, and the male has a melodious, high-noted song during the breeding season. Silvereyes are omnivores but enjoy fruit, nectar and seeds.

## Swallow (Family Hirundinidae)

The **Welcome Swallow** *Hirundo neoxena* is also a recent arrival, self-introduced from Australia during the 1950s, and is now common on open land, especially around lakes, rivers and ponds. It often builds its nests of mud and grass in corners and on the rafters of buildings. These are small birds (10–20g) and their forked tail and fast flying make them noticeable, along with their rust-coloured head, neck and breast. Their backs and upper wings are black and their undersides are creamy.

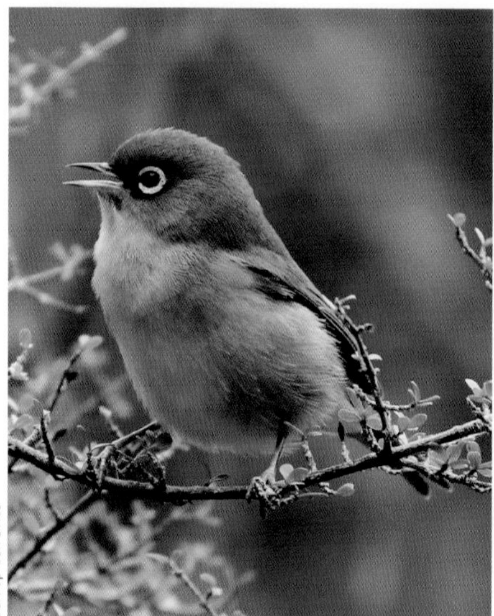

Silvereye

Welcome Swallow

# Glossary of Māori and English Bird Names

Because birds are often known locally in New Zealand by their Māori names, these have generally been included alongside English common names. The following list gives the Māori names and their English equivalents, as well as the scientific names. Bird names marked with an asterisk indicate that the Māori name only is used in the book.

| Māori name | English name | Scientific name |
|---|---|---|
| hīhī | Stitchbird | *Notiomystis cincta* |
| hoiho | Yellow-eyed Penguin | *Megadyptes antipodes* |
| *kākā | Kaka | *Nestor meridionalis* |
| *kākāpō | Kakapo | *Strigops habroptilus* |
| kākāriki | Red-crowned Parakeet | *Cyanoramphus novaezelandiae* |
| kākāriki | Yellow-crowned Parakeet | *Cyanoramphus auriceps* |
| kakī | Black Stilt | *Himantopus novaezelandiae* |
| kārearea | New Zealand Falcon | *Falco novaeseelandiae* |
| karoro | Southern Black-backed Gull | *Larus dominicanus* |
| *kea | Kea | *Nestor notabilis* |
| kererū | New Zealand Pigeon | *Hemiphaga novaeseelandiae* |
| *kōkako | Kokako | *Callaeas wilsoni* |
| korimako | Bellbird | *Anthornis melanura* |
| kororā | Little Penguin | *Eudyptula minor* |
| kōtuku | White Heron | *Ardea alba* |
| mātātā | Fernbird | *Bowdleria punctata* |
| mohua | Yellowhead | *Mohoua ochrocephala* |
| parea | Chatham Island Pigeon | *Hemiphaga chathamensis* |
| pāteke | Brown Teal | *Anas chlorotis* |
| pīpipi | Brown Creeper | *Mohoua novaeseelandiae* |
| pīwakawaka | New Zealand Fantail | *Rhipidura fuliginosa* |
| poaka | Pied Stilt | *Himantopus leucocephalus* |
| *pūkeko | Purple Swamphen | *Porphyrio melanotus* |
| roaroa | Great Spotted Kiwi | *Apteryx haastii* |
| rowi | Okarito Brown Kiwi | *Apteryx rowi* |
| ruru | Morepork | *Ninox novaeseelandiae* |
| tāiko | Westland Petrel | *Procellaria westlandica* |
| *takahē | Takahe | *Porphyrio hochstetteri* |
| tara | Black-fronted Tern | *Chlidonias albostriatus* |
| tarāpuka | Black-billed Gull | *Chroicocephalus bulleri* |
| tauhou | Silvereye | *Zosterops lateralis* |
| tawaki | Fiordland Crested Penguin | *Eudyptes pachyrhynchus* |
| tīeke | North Island Saddleback | *Philesturnus rufusater* |
| tīeke | South Island Saddleback | *Philesturnus carunculatus* |
| tītī | Sooty Shearwater (Muttonbird) | *Puffinus grisea* |
| tokoeka | South Island Brown Kiwi | *Apteryx australis* |
| tōrea | South Island Pied Oystercatcher | *Haemotopus finschi* |
| toutouwai | South Island Robin | *Petroica australis* |
| *tūī | Tui | *Prosthemadera novaeseelandiae* |
| turiwhatu | Banded Dotterel | *Charadrius obscurus* |
| *weka | Weka | *Gallirallus australis* |
| whio | Blue Duck | *Hymenolaimus malacorhynchos* |

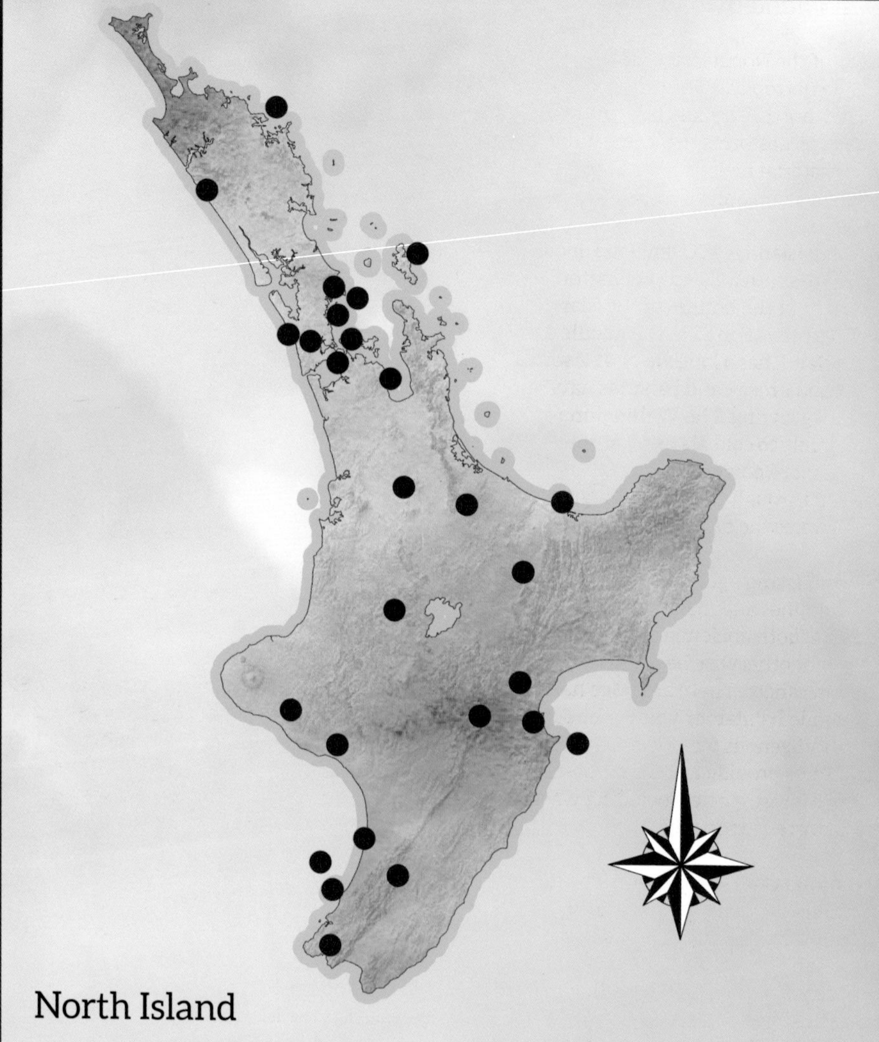

# North Island

1 Bay of Islands, Iripiri Project Island Song
2 Trounson Kauri Park
3 Aotea/Great Barrier Island
4 Tāwharanui Regional Park
5 Tiritiri Matangi
6 Shakespear Regional Park
7 Muriwai Gannet Colony
8 Western Springs
9 Ambury Regional Park and Watercare Coastal Walkway
10 Motutapu Island
11 Pūkorokoro Miranda Shorebird Centre
12 Sanctuary Mountain, Maungatautari
13 Sulphur Bay Wildlife Refuge, Rotorua
14 Ohiwa Harbour
15 Whirinaki Te Pua-a-Tāne Conservation Park
16 Pureora Forest Park
17 Lake Rotokare Scenic Reserve.
18 Bushy Park Sanctuary, Whanganui
19 Manawatū Estuary
20 Ahuriri Estuary, Napier
21 Boundary Stream and Shine Falls
22 Blowhard Bush Scenic Reserve
23 Cape Kidnappers Gannet Colonies
24 Pūkaha Mount Bruce National Wildlife Centre
25 Waikanae Estuary
26 Kāpiti Island Nature Reserve
27 Zealandia Eco-sanctuary

## North Island overview

The top of the North Island, at latitude 34.5 south, is subtropical. One can grow bananas in the far north. The climate is influenced by subtropical storms that come down from equatorial regions bringing warm, thunderous rain and high winds.

The North Island is 1,100km long, mostly on a north–south axis, so Wellington, the capital city, at the bottom of the island, and democratically located in the middle of New Zealand, has a latitude of 41.2 south. It's cool and breezy and bananas have no chance of growing. The Wellington region is susceptible to cold blasts of Antarctic air made more powerful by being tightly funnelled through the Cook Straight, the sea that separates the two islands.

The North Island's geography is complex with fault lines and massive areas of volcanism, both ancient and active, upsetting what might otherwise be a clearer pattern to mountain ranges. There are huge harbours that provide habitat for wading birds. Remote indigenous forests that escaped the millers' saws provide habitat for passerines and other forest species including whio, an unusual torrent-living duck.

The Hauraki Gulf, with its many predator-free islands, is a global seabird hotspot, with the added advantage to tourists that it lies on the doorstep of Auckland, the country's largest city. It supports 27 breeding seabird species, five of which breed only in the Gulf. Hauturu/Little Barrier Island, in the heart of the Gulf, is a 28 square-kilometre, forest-covered, ancient volcano and a predator-free haven for many species of indigenous birds. It's a closed bird sanctuary and one can't visit but it is good to know it's there. Numerous endangered species thrive on this island, with some populations becoming strong enough for translocations to other areas that have, more recently, become predator-free.

There are some bird species that are found in the North Island but not the South, the kōkako and the hīhī, for instance, and vice versa. Also, the islands are separated enough by sea for some species to have evolved separate North and South Island subspecies.

### Iconic bird

The kererū, a large, handsome pigeon, white-breasted with dark, iridescent wings and top sides, is admired for its beauty and for its acrobatic flying displays. It's found throughout New Zealand, including in some urban areas. Kererū is an important forest vector; it eats fruits, leaves, flowers and, as it often flies long distances, it spreads seeds helping with the regeneration of indigenous forest.

Liz Light

*Motutapu Island, Hauraki Gulf; a classic combination of pastureland, forest and beaches*

# Bay of Islands, Iripiri Project Island Birdsong

Dean Wright

*Urupukapuka Island, one of seven predator-free islands*

In 2009 seven islands in the eastern Bay of Islands, and all the little islets surrounding them, became predator free. Attaining this was a massive community project. Since then 130,000 native trees and shrubs, across 60 different species, have been planted to restore the habitat to how it once was and to provide food for a bigger variety of birds. Recently five species of bird have been translocated to the islands. The North Island Robin, pāteke/Brown Teal, tīeke/North Island Saddleback, kākāriki and Whitehead are now thriving, along with the native species that had survived the onslaught of introduced predators and farming. Nine more species, once present on the

Liz Light

*Pastureland, forest and seashore provide a varied habitat for numerous bird species*

## KEY FACTS

### Getting There

Explore run a ferry service to Otehei Bay on Urupukapuka Island. There are four daily return journeys from 1 October to 28 April, and more in December and January. Ferry timetable, see: https://russellinfo.co.nz/timetable-otehei-bay. Costs: adult $45 return, child $25. There is also a hop-on, hop-off ferry service that goes to Moturua and Motuarohia Islands, but rather than spending a shorter time at all the islands, it is better to spend the day on Urupukapuka and walk the tracks (www.exploregroup.co.nz/destinations/bay-of-islands/walk/island-escape-hop-on-hop-off).

### Facilities

Urupukapuka has a cafe/bar in Otehei Bay (open 9 a.m. to 5 p.m.) and a conservation centre that gives lots of information about Project Birdsong and the islands. There are toilets in Otehei Bay and the camping areas.

### Habitat

Pastureland, replanted forest, established forest, wetlands and beaches.

### Best Time to Visit

There is good birding in all seasons but more bird activity and song in spring. This area is quite sub-tropical so even in winter there are lovely days.

islands, will be reintroduced in the next few years. Urupukapuka, the largest island, has numerous white sand beaches with crystal clear water, delightful walks with good birding and amazing views. These islands are a paradise for people as well as for birds. See www.projectislandsong.co.nz.

## Tracks and walks

There are numerous tracks and walks of different lengths. Visit the conservation centre in Otehei Bay, where the ferry arrives, and get the latest information on the best birding walks. Walking map: www.exploregroup.co.nz/media/1303/walking-map.jpg.

Behind Otehei Bay, check out the wetlands for herons, poaka/Pied Stilts, Paradise Shelduck, pūkeko and other bird visitors. Walk up the hill (great views of the Bay of Islands from the ridge) and along the ridge through mānuka and kānuka forest. Here you will probably see small forest birds such as Whitehead, pīwakawaka/New Zealand Fantail, North Island Robin, and flocks of tauhou/Silvereye twittering higher in the trees. Some of these birds are not shy and will come close to eat insects raised from your footfalls in the leaf litter.

### Key Species

Tīeke/North Island Saddleback, Whitehead, North Island Robin, kākāriki/Red-crowned Parakeet, pāteke/Brown Teal, North Island Kiwi, kererū/New Zealand Pigeon, North Island Tomtit, pīwakawaka/New Zealand Fantail, New Zealand Pipit, Grey Warbler, tūī, Paradise Shelduck, Fernbird, Swamp Harrier, Banded Rail, Sacred Kingfisher, pūkeko, ruru/Morepork, tauhou/Silvereye, Welcome Swallow, Shining Cuckoo, kororā/Little Penguin, New Zealand Dotterel, Southern Black-backed Gull, Pied Shag, Australasian Gannet, Reef Heron, White-faced Heron, Variable Oystercatcher, Spur-winged Plover, Grey-faced Petrel.

Follow the Pāteke Loop and find a hide alongside a small lake/wetland, where you will see pāteke/Brown Teal and other waterbirds. Then, at Otiao Bay, watch out for shorebirds such as oystercatchers and New Zealand Dotterels. Most people from the ferry never get this far so you will probably have these tracks to yourself. This can be done in three hours but make a day of it. Carry plenty of water and food.

*Pāteke*

Dean Wright

*North Island Robin*

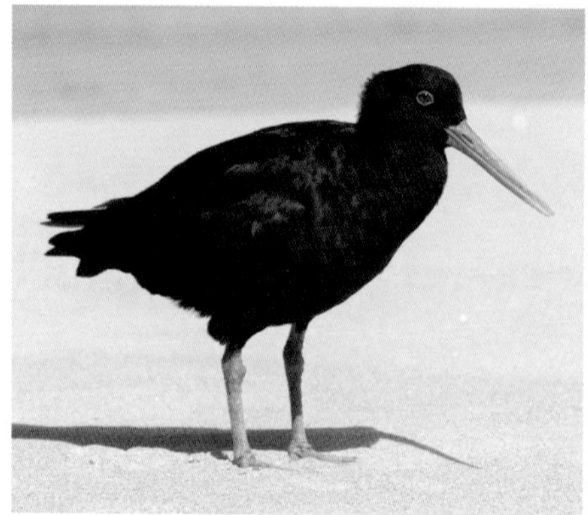

Gert op den Dries

## Variable Oystercatcher

*Haematopus unicolor*

Oystercatchers inhabit beaches, dunes and sandspits, and are not migratory. They are monogamous and territorial, often staying in the same areas and stridently defending their territories from people, dogs and other intruders, especially when they are breeding. They are glossy black, and have a long red bill and red eyes. Being attacked by this screeching creature, glowing red sword to the fore, is positively scary. The trapping of predators, which eat chicks and eggs, has benefited this species. Its numbers are steadily increasing with, now, an estimated population of 5,000 or more. The birds are found on most of the sandy beaches on Urupukapuka Island.

### Tūī

*Prosthemadera novaeseelandiae*

Tūī are found in all of the North Island and much of the South Island, and are an iconic songbird. They combine bell-like notes with rasps, creaks and thumps, and in many areas this is the sound that wakes people in the morning. They are dark – black, navy blue and brown – with a petrol-on-a-puddle iridescent sheen. Tūī have a tuft of white feathers on the throat, hence the sometimes name parson bird. They are aggressive defenders of their territory and nests, and big enough (30cm long) to mean business. They are nectar feeders by preference but go for fruit in autumn. The birds are found in many locations on Urupukapuka Island and, as nectar feeders, they can be seen where trees, flaxes and shrubs are flowering or fruiting. Their morning and evening songs are loud and melodious.

Liz Light

## Accommodation

There are three Department of Conservation (DOC) campsites on Urupukapuka Island, all in stunning white-sand bays. See the walking tracks map mentioned on page 60. Unless you have your own boat it is a haul to get gear from Otehei Bay wharf, where the ferry berths, to the campsites. A water taxi is a better option (www.boiwatertaxi.co.nz).

# Trounson Kauri Park

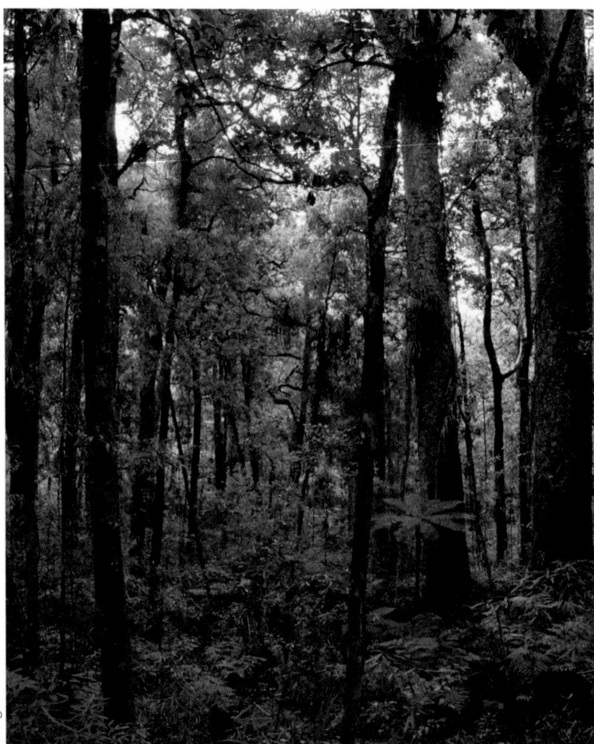

Liz Light

*Trounson Kauri Park features thousands of tall, straight, kauri trees*

## KEY FACTS

### Getting There

Trounson Kauri Park is 38km north of the town of Dargaville on the west coast of Northland. It is accessible via Trounson Park Road from the south off State Highway 12 or from the north via Katui Road, which links to Trounson Park Road at Donnellys Crossing. There are two entry points to the park; the Trounson Kauri Park day car park and the Trounson Kauri Park campground.

### Facilities

The nearest petrol station is 9km south at Kaihu, and 37km north at Waimamaku. There is no shop or cafe so bring your own food and water.

### Habitat

Kauri forest and adjoining pastureland.

### Best Time to Visit

There is good birding in all seasons but more bird activity and song in spring. This area is quite subtropical so even in winter there are lovely days.

### Useful Website

www.doc.govt.nz/parks-and-recreation/places-to-go/northland/places/trounson-kauri-park.

Trounson Kauri Park is a 450ha mainland island. This means that predatory pests are trapped and consequently bird life is abundant. The park is special because of the flora. Here, as its name implies, kauri trees are dominant. Kauri only grow in the north of New Zealand and the biggest of them rank as some of the world's largest trees, with some being more than 2,000 years old. The forest giants are tall and straight; magnificent symmetrical columns with vast, spreading crowns. They grow in groups in tall stands. Trounson Kauri Park has fine examples of mighty kauri. Some birds live high in the canopy and are more often heard than seen. Others, such as the pīwakawaka/New Zealand Fantail, Whitehead, tauhou/Silvereye and Grey Warbler, tend to live in the lower reaches of the forest and on the forest edges. There are guided night walks and Northern Brown Kiwi are commonly seen. If they are not sighted, their calls and their feet shuffling around in the understorey can be heard. The park is a few kilometres off the main road north and not many people visit, but it is a treat for both its flora and fauna.

## Tracks and walks

There is a **loop track** of 1.6km. It is well paved and has boardwalks over low-lying and damp parts. It passes through spectacular kauri glades, past a stream and through dells of filmy ferns where the sun shines into the understorey. It is imperative to stay on the track at all times to ensure that no damage is done to the sensitive roots of the magnificent kauri trees, and to help prevent the spread of kauri dieback disease.

### Key Species

Northern Brown Kiwi, Black Shag, Little Shag, Paradise Shelduck, Grey Duck, Australasian Harrier, North Island Weka, Spur-winged Plover, pūkeko, kererū/New Zealand Pigeon, kākā, Long-tailed Cuckoo, Shining Cuckoo, kōkako, ruru/Morepork, Sacred Kingfisher, Grey Warbler, Welcome Swallow, korimako/Bellbird, tūī, pīwakawaka/New Zealand Fantail, New Zealand Pipit, Whitehead, tauhou/Silvereye.

Liz Light

*Ferns in Trounson Kauri Park*

Gert op den Dries

*Silvereye*

## Northern Brown Kiwi

*Apteryx mantelli*

This bird is flightless, nocturnal, and has whiskers, nostrils at the end of its long beak and long, strong legs. It has dark brown feathers streaked with reddish and black shades. The female lays one huge egg annually, the largest egg in the world in proportion to body size; some 20 per cent of it. Kiwi nest in burrows, holes in trees and other sheltered cavities. The chicks hatch fully formed and only stay in the nest for five days, then are out foraging with their parents. Kiwi can live for 50 years and grow to 25–40cm tall. Being nocturnal, they are heard more than they are seen and the male has a trilling, warbling cry with rising notes. The female is quieter and gruffer. The birds sometimes duet. The Northern Brown Kiwi is the most common kiwi species though because of habitat loss, and predation by dogs, cats and stoats, the population, estimated to be up to 21,000, is declining in areas without predator control. It is thriving in the sanctuary, can be heard from the camping ground at night and is often seen on night walks.

Oscar Thomas

## Shining Cuckoo

*Chrysococcyx lucidus*

The Shining Cuckoo nests in New Zealand but migrates to northern Pacific countries in the winter. Its loud, distinctive and repetitive whistle is the avian announcement of spring's arrival. It is medium sized and has bronzed green wings, grey-and-white striped breast and underparts, and a black crown. It is shy and well camouflaged, but can be seen if birdwatchers concentrate on following the call to where it is perched. The species is a nest parasite and lays one egg in the nest of other smaller birds, usually the tiny Grey Warbler (see page 80). The warbler hatches the egg, then feeds what becomes a ridiculously big chick in comparison to the warbler. The busy little warblers are adapted to this situation and they usually rear a clutch of their own chicks before the cuckoo arrives in New Zealand, in spring.

## Accommodation

There is a small but very sweet camping ground with shower, toilets and kitchen amenities. It has eight powered sites, primarily for camper vans, and 12 non-powered sites for tents. It does not have a booking system. Those who arrive first get the places but it is seldom full.

# Aotea/Great Barrier Island

*Beach, estuarine and forest land*

This place, two hours by ferry from Auckland (or 25 minutes in a small plane), has a population of 800, a few little towns, a couple of stores and petrol pumps, and kilometres of beautiful beaches and bush. It is a remote paradise for people and not quite yet a paradise for birds. The island is free of possums, stoats, ferrets, weasels, hedgehogs, deer and feral goats. Predators that have plagued the mainland forests for 150 years were never established here. However, rats, feral cats and wild pigs are a continuing problem and have a devastating effect on the native birds. DOC is steward of 12,000ha, 70 per cent of this island, and it is possible that before long a concerted effort will be made to get rid of the remaining predators.
In the island's centre, spectacular bluffs and ridges rise to Hirakimata/Mt Hobson (621m). To the west, forest-covered ranges meet the coast, in a maze of bays, islands and indented fiords. The eastern coastline has sweeping white-sand surf beaches, often backed by tidal creeks and wetlands. These locations make for a variety of different habitats for birds.

Liz Light

*Kaitoke Swamp*

## KEY FACTS

### Getting There
Great Barrier Island lies 100km north-east of Auckland on the outer edge of the Hauraki Gulf. Several companies service the island by sea and air. Small plane flights leave from four mainland airfields, and there are both passenger and vehicle ferry services. Book through the 'getting there' links on: www.greatbarrierislandtourism.co.nz.

### Getting around
This is a big island and you will need transport. Taxis and car rentals are available on the island. Great Barrier Island Rentals offers numerous car options at reasonable prices. It will arrange to meet you on arrival at the Claris Airfield or Tryphena Wharf with a pre-booked vehicle, an island map and a rental agreement. Bookings essential (www.greatbarrierislandtourism.co.nz/rental-cars-and-vans).

### Habitat
This is a large island with many habitats, including sandy beaches, rocky headlands and cliffs, swamp, rivers, pastureland, kauri forest and scrubland.

### Best Time to Visit
Summer holidays (January and February) and Easter can be busy with hikers, beach-goers and campers from Auckland, so it is difficult to get accommodation and hire a car. July and August can be cold and muddy. Many of the birds are resident, so all other seasons are good.

### Useful Website
www.greatbarrierislandtourism.co.nz.

## Tracks and walks

There are many tracks and birdwatching sites. It depends on what you are looking for and how much time you have.

Seabirds are plentiful. The Black Petrel breed on the island, as does Cook's Petrel. Grey-faced Petrels, Fluttering Shearwaters and Diving Petrels breed on some of the nearby islets. Little Penguins and gannets are common. Other seabirds that breed on predator-free islands in the Hauraki Gulf are often seen. To see seabirds easily it is best to have access to a boat. There are fishing charter boats on the island that can easily be diverted from fishing to seabird watching.

Shorebirds include the Wrybill, and New Zealand and Banded Dotterels. Migratory waders like Bar-tailed Godwits and Red Knots are present on the Okiwi Spit, in season. Kaitoke Swamp, a large wetland, has a big population of Fernbirds as well as some Banded Rail, and Spotless Crake. Pāteke/Brown Teal are common and can be seen around the stream at Tryphena village, in

### Key Species

Kororā/Little Penguin, White-faced Storm Petrel, Common Diving Petrel, Black-bellied Storm Petrel, Grey-backed Storm Petrel, New Zealand Storm Petrel, Flesh-footed Shearwater, Buller's Shearwater, Fluttering Shearwater, tītī/Sooty Shearwater, Short-tailed Shearwater, Grey Petrel, Little Shearwater, White-headed Petrel, Black Petrel, Northern Giant Petrel, Cook's Petrel, Black-winged Petrel, Mottled Petrel, Broad-billed Prion, Slender-billed Prion, Australasian Gannet, Black Shag, Pied Shag, Little Shag, Paradise Shelduck, pāteke/Brown Teal, White-faced Heron, Reef Heron, Australasian Harrier, Banded Rail, North Island Weka, Pied Oystercatcher, Variable Oystercatcher, Spur-winged Plover, pūkeko, Spotless Crake, New Zealand Dotterel, Banded Dotterel, Golden Plover, Eastern Bar-tailed Godwit, Arctic Skua, Southern Black-backed Gull, Red-billed Gull, Caspian Tern, White-fronted Tern, poaka/Pied Stilt, kererū/New Zealand Pigeon, kākā, kākāriki/Yellow-crowned Parakeet, Long-tailed Cuckoo, Shining Cuckoo, ruru/Morepork, Sacred Kingfisher, Grey Warbler, Tomtit, North Island Robin, Welcome Swallow, korimako/Bellbird, tūī, pīwakawaka/New Zealand Fantail, New Zealand Pipit, Whitehead, Fernbird, tauhou/Silvereye.

Glenda Peake

*Pied Shag*

the wetlands of Whangapoua Harbour and in many other wetland areas. Kākā are often seen near the towns, especially around Port FitzRoy and Tryphena.

**Glenfern Sanctuary**, overlooking FitzRoy Harbour, provides an easy way to see a variety of bird species. A predator-proof fence built in 2009 across the Kōtuku Peninsula, and the subsequent eradication of rats, pigs and cats, keep the area totally predator free and the birds have thrived. There are easy walks through different habitats, and accommodation on the property and nearby. Glenfern also takes bookings and organizes guided walks with naturalists on other parts of the island (www.glenfern.org.nz).

The seas inside Great Barrier, in the Hauraki Gulf, have one of the world's highest diversities of petrels and shearwaters, with 13 breeding species. At least five of these breed on Great Barrier and surrounding islets, and most of them may be seen at sea around the island.

## Tāiko/Black Petrel

*Procellaria parkinsoni*

There is a nesting colony of more than 1,000 pairs of tāiko around the summit of Hirakimata/Mt Hobson and other high points in the centre of the island. There are also 100 or so pairs on nearby Hauturu/ Little Barrier Island. Tāiko are endemic and these are the only known colonies. Tāiko are medium-sized petrels and fully black but for a yellowish beak. They are monogamous, usually for life. Both parents incubate the egg and care for the chick. The chicks fledge at 96–122 days old. After a chick fledges it flies solo to tropical South America and the eastern Pacific, and remains there until the breeding season comes around. If you stay in the Mt Heale hut during the summer months, you can hear and see the birds returning to their burrows at night. In the breeding season tāiko are often seen fishing in the outer Hauraki Gulf and, because they night fish, primarily for luminous squid,

Oscar Thomas

they are severely at risk from commercial fishing. New Zealand fishers have a range of compulsory fishing techniques that they use to mitigate against catching them.

## Pāteke/Brown Teal

*Anas chlorotis*

The endemic pāteke lives in lowland swamps, streams and ponds, and estuaries. The birds enjoy foraging on beaches near stream outlets and are often seen near the stream at Tryphena Beach. The island has the largest population of pāteke, but the birds are at risk from domestic pets, especially cats and dogs. The trapping of predators and habitat awareness has taken the pāteke from being listed as endangered to recovering. There have been numerous successful translocations to predator-free sanctuaries. Pāteke are mottled brown, monogamous and territorial, and usually lay a clutch of 4–5 eggs. They are omnivorous and eat a wide variety of grasses, seeds, insects, molluscs, worms and many other small invertebrates.

Liz Light

## Kākā

*Nestor meridionalis*

Kākā, a forest bird in both the North and South Islands, is a charming, rowdy and colourful parrot. Kākā are medium-sized (450g), olive-brown parrots with glamorous red-orange underwings. They are often heard before they are seen, uttering a loud *ka kaa* cry as they fly high in the sky. They do well in forest areas and the edges of forests where mammalian pests have been eliminated or are controlled. The birds are mainly seed, fruit and nectar eaters, with excellent memories of the location of seasonal food sources. They are common on Aotea/Great Barrier Island, where they swing on clothes lines, eat flax nectar and raid the locals' orchards.

Liz Light

*The pastureland and wetlands of Harataonga Bay*

## Accommodation

There are numerous accommodation options, including cottages, motels and lodges, bed and breakfasts, and nine camping grounds (www.greatbarrierislandtourism.co.nz/accommodation).

# Tāwharanui Regional Park

*Tāwharanui Regional Park*

## KEY FACTS

**Getting There**
The park is 90km north of Auckland by car with a journey time of 1½ hours. There is no public transport. Stop at the interpretation area just inside the sanctuary gate for information. Near the lagoon is a great place for ducks, shags and other waterbirds. Access to the park is free. Take all your food and drink with you and remove all your rubbish.

**Habitat**
Native forest, coastal cliffs, sandy beaches, brackish lake, wetlands and pastureland.

**Best Time to Visit**
Summer holidays (January and early February) can be busy with beachgoers and campers. July and August can be cold and muddy. Many of the birds are resident so all other seasons are good.

**Useful Website**
www.tossi.org.nz. On the the 'park' tab you can download an excellent birders' information PDF, including a map and bird list.

Tāwharanui Regional Park is on a finger of land stretching into Hauraki Gulf 90km north of Auckland. There are pretty beaches on the north side and steep forested cliffs on the south side. In 2003 a predator-proof fence was built across the bottom of the peninsula and Auckland Council started a concentrated campaign to rid the park of introduced animals that had been decimating the native birds for 200 years. In 2005 Tāwharanui became predator free. The park is known as an integrated open sanctuary and combines farming, recreation and conservation of native species. There is a variety of bird habitats, including brackish ponds for waterbirds, wetlands for margin dwellers, beaches for shorebirds, rainforest for kiwi and passerines, and rocky coastland for seabirds. It is also a beautiful place for swimming, camping, surfing and walking.

*Tāwharanui features a variety of different habitats and bird species*

## Tracks and walks

**The lagoon**, near the park gate, is a great place for spotting the pāteke/Brown Teal, Paradise Shelduck, Pied Shag, Shoveler, Banded Rail, Variable Oystercatcher, poaka/Pied Stilt and other species, depending on the day. The nearby wetland is good habitat for the takahē, kiwi (at night), bittern and pūkeko.

There are seven signposted walks in the park. They begin at Anchor Bay and the lagoon, and range from easy strolls to the two-hour **Coast Track**. The end of the peninsula is a great place for spotting seabirds. Grey-faced Petrels, Fluttering Shearwaters and Diving Petrels nest here in spring and summer. The sea adjoining the north coast is a marine reserve so is a rich feeding ground for seabirds. Australasian Gannets can often be seen diving here. New Zealand Pipits are found on the central farmland tracks.

A favourite walk is the **Ecology Trail** (two hours) signposted from Anchor Bay. In this rainforest valley there are glades of nīkau palms, ponga (tree ferns) with lacy leaves, giant pūruri trees and many other indigenous tree species. Here korimako/Bellbirds chime, pīwakawaka/New Zealand Fantails flit across the path, robins hop about, kererū/ New Zealand Piegeons feast on flowers and tūī chase each other, weaving through trees at speed. There are kiwi but they do not present themselves to day-trippers.

The north-facing beaches have numerous resident dotterels and Little Penguins can be seen at dusk, in spring and summer, coming from the sea and across the beach to their nests.

Glenda Peake

*Banded Rail*

### Key Species

**Terrestrial:** Northern Brown Kiwi, korimako/ Bellbird, Grey Warbler, pīwakawaka/Fantail, kākā, kākāriki/Red-crowned Parakeet, Whitehead, North Island Robin, kererū/ New Zealand Pigeon, tūī, New Zealand Pipit, Paradise Shelduck, pāteke/Brown Teal, Banded Rail, Sacred Kingfisher, Australasian Harrier, Shining Cuckoo, Long-tailed Cuckoo, ruru/ Morepork, pūkeko, tauhou/Silvereye, Spotless Crake, Welcome Swallow, Australasian Shoveler, Grey Duck, Australasian Bittern, takahē.
**Shorebirds and seabirds:** Red-billed Gull, Black-backed Gull, Reef Heron, White-faced Heron, Variable Oystercatcher, kororā/Little Penguin, Common Diving-petrel, Grey-faced Petrel, Pied Shag, Buller's Shearwater, Flesh-footed Shearwater, Arctic Skua, Spur-winged Plover, Caspian Tern, White-fronted Tern, New Zealand Dotterel. Numerous other shearwaters and petrels other than those mentioned above can be seen at sea in season.

Oscar Thomas

### New Zealand Dotterel

*Charadrius obscurus*

The New Zealand Dotterel is an endemic beach bird with, in breeding plumage, a peachy chest and spangled brown, cream and grey topsides. There are 17 breeding pairs on two beaches within the park and they can be easily be seen in most seasons. They thrive because the park is predator ree. These are endearing little beach birds, vith chicks like fluffy pingpong balls on ast, twiggy legs. Their nests are scrapes in he sand above the tide line. There are only ıbout 2,000 birds left. In other places the ›opulation struggles because predators, ·specially cats and stoats, prey upon adults, ·leeping on their nests at night, and on their :iny chicks.

## Korimako/Bellbird

*Anthornis melanura*

The Bellbird is a small endemic bird with olive-green colour variations on most of its body but dark (black/purplish) tail and wings. It is primarily a nectar eater but also enjoys fruit and insects. The birds are not special to look at but their song, usually that of the male, is superb; the bell-like notes are clear and pure and they produce great pitch and tonal variation for such small birds. They often gather in groups, fluff themselves up and sing in a heavenly chorus. In 2005 a flock of 100 or so Bellbirds flew over from Hauturu/Little Barrier Island and made Tāwharanui their home. Their song fills the bush on the Ecology Trail and they are now the most common bird in the park.

## Accommodation

Book a campsite, or a stay in a rental cottage (www.aucklandcouncil.govt.nz/parks-recreation/stay-at-park). It is a beautiful camping area. Besides great birding there are two gorgeous beaches so you can swim, surf or snorkel, or dive in the Marine Protected Area on the north coast. It is wonderful to stay in the park at night to go kiwi spotting and hear the ruru call. There are numerous bed and breakfast accommodations nearby. See Airbnb or Booking.com.

# Tiritiri Matangi

Liz Light

*Tiririri Matangi's steep east coast bays*

## KEY FACTS

### Getting There

There is a ferry from Auckland ferry terminal five days a week (not Monday and Tuesday) and on public holidays. It departs from Auckland at 9 a.m. and leaves the island at 3.30 p.m. The journey, which is scenic, takes one hour. See the website for information (www.360discovery.co.nz).

### Facilities

The visitor centre provides free coffee, tea and water – otherwise you must take your own food and drink. There are toilets, a small eco-shop and picnic tables. There are beautiful places to picnic on the island. Swimming is safe on the west side though the beaches are stony.

### Fees

There is no fee to visit this bird sanctuary but you must pay the ferry ticket. It is also possible to visit by private boat. A guided walk costs $10.00 (children $2.50). The money from guided walks helps fund conservation work.

### Best Time to Visit

Any season is good. Early spring, in September, when the kōwhai and clematis are in flower, and many bird species are courting, is lovely. In summer you can see bird families and chick feeding.

### Useful Website

www.tiritirimatangi.org.nz.

Tiritiri Matangi is a lozenge-shaped island bird sanctuary in the Hauraki Gulf, 40km from Auckland. It is easily accessible by ferry. The island is a 220ha emerald gem surrounded by shining blue sea. The ferry journey is pleasant and there are birdwatching walks for every fitness level. Twenty-four species of endemic and native terrestrial birds are resident on the island, as well as 15 shorebirds and seabirds.

A lighthouse was built on the island in 1864 and the native bush was mostly destroyed to make the lighthouse keeper's farm. By 1984 the farm was uneconomic and the lighthouse was automated so the area became a conservation reserve. More than 280,000 native trees were planted, and when the planting programme finished the focus turned to

Liz Light

*The morning sun catches the golden heads of the cabbage trees on the Wattle Track*

repopulating the island with bird species that would have been there before the farming years, as well as others that needed a safe haven. In 1993 kiore rats, the last of the introduced pests, were exterminated, and the bush and the birds have thrived ever since. Numerous species have bred so abundantly that translocation of birds to other safe havens has been possible. Tiritiri Matangi is an outstanding conservation success story and a prime place for birdwatchers. Many rare species can be seen here throughout the year.

### Key Species

**Endemic:** Little Spotted Kiwi, korimako/Bellbird,Grey Warbler, pīwakawaka (New Zealand Fantail), takahē, kākāriki/Red-crowned Parakeet, Whitehead, North Island Robin, hīhī/ Stitchbird, kōkako, tīeke/North Island Saddleback, Fernbird, Rifleman, kererū/New Zealand Pigeon, Paradise Shelduck, tūī.
**Native:** Australasian Harrier, Shining Cuckoo, Kingfisher, ruru/Morepork, pūkeko, tauhou/Silvereye, Spotless Crake, Welcome Swallow.
**Shorebirds and seabirds:** Red-billed Gull, Black-backed Gull, Reef Heron, Variable Oystercatcher, kororā/Little Penguin, Common Diving-petrel, Grey-faced Petrel, Pied Shag, Buller's Shearwater, Flesh-footed Shearwater, Arctic Skua, Spur-winged Plover, Caspian Tern, White-fronted Tern.

## Tracks and walks

See PDF link to map: www.doc.govt.nz/Documents/parks-and-recreation/places-to-visit/auckland/tiritiri-matangi-map.pdf.

**Ridge Road** follows the central ridge the length of the island and there are tracks to the east and west of it, linking up, to circle the island. Several forest paths link these three tracks so there are plenty of options. Walking the exterior tracks takes about four hours. The walk from the wharf to the visitor centre is 1km. The two tracks used for the guided walks are about 2km (Wattle Track) and 3km (Kawerau Track).

Tūī, korimako/Bellbirds, kererū/New Zealand Pigeons and pīwakawaka/Fantails are common and can be spotted on most tracks. Pīwakawaka often follow walkers, flitting

*North Island Saddleback*

around and cheeping. The **Wattle Track** is good for spotting hīhī/Stitchbirds and korimako/Bellbirds because there are sugar feeders near it. There are gorgeous views across the Gulf islands towards Auckland. An abundance of cabbage trees, *Cordyline australis*, an unusual tree endemic to New Zealand, makes this walk special. Pūkeko and takahē are primarily grass eaters so they are found on pastureland, especially near the visitor centre. North Island Saddlebacks (tīeke) are often to be seen scratching around in leaf litter in forested areas, looking for insects, on **Hobb's, Totara and Wattle Tracks**. Wattle Valley is home to a pair of North Island kōkako that have reared many chicks over the years. These large blue wattlebirds have a haunting, flute-like call.

There are plenty of shorebirds and seabirds, in season. There are breeding colonies of Red-billed and Black-backed Gulls, and Reef Herons are often seen. Several pairs of Variable Oystercatchers nest in the wharf area and kororā/Little Penguins nest near the shore. There are thriving colonies of Common Diving-petrels, Grey-faced Petrels and Pied Shags. Caspian and White-fronted Terns are often seen and nest on the island.

Oscar Thomas

*Little Penguin*

Oscar Thomas

*Black-backed Gull*

Oscar Thomas

## Takahē

*Porphyrio hochstetteri*

Takahē, the world's largest rail (up to 2.7kg), has spectacular colouring with a dark iridescent blue breast and underparts, forest-green back and wings, red legs and chunky, V-shaped bright red beak. Its wings are remnants of eons of evolutionary history and it is flightless. Takahē were thought to be extinct until a few were rediscovered in a remote South Island mountain valley in 1948. Captive breeding and release to predator-free sanctuaries has saved this species. There are now more than 300 birds and the number is increasing by 10 per cent each year. Tiri has played its part in the successful breeding of takahē and translocation to other safe havens. Some takahē are habituated to humans and will try to scrounge your sandwiches. Do not feed them. Grasses and fern rhizomes are their primary food, so they tend to stay near grassland by the lighthouse. They are so big that you cannot miss them.

## Hīhī/Stitchbird

*Notiomystis cincta*

Hīhī were initially categorized as honeyeaters but now have a taxonomic category of their own. Before European settlement, this medium-sized forest passerine lived in much of the North Island. It was prone to predation and suffered more than most birds from avian diseases and parasites that arrived along with introduced birds. By 1980 there were no surviving hīhī on the mainland and one remaining population on Hauturu/Little Barrier Island, where there are now some 3,000 birds. As a result of good conservation management and appropriate translocations, hīhī are thriving in seven predator-free bird sanctuaries. At Tiritiri Matangi there are supplementary sugar-feeding stations to help the birds survive the low-nectar seasons. They are easy to see at these stations, particularly on the Wattle Track. See also www.hihiconservation.com.

## Accommodation

It is possible to stay in the DOC bunkhouse for a fee. Book online at DOC website (www.doc.govt.nz/tiritiribunkhouse). Beds are often booked several months ahead, especially in summer. It is a superb experience to go kiwi spotting at night and to be there for the dawn chorus.

# Shakespear Regional Park

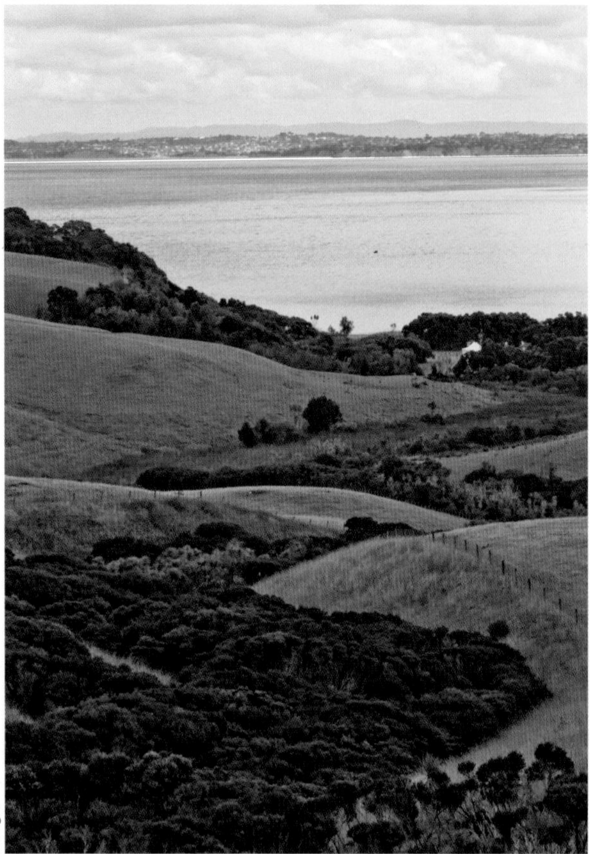

Liz Light

*Shakespear Regional Park features a variety of different bird habitats*

## KEY FACTS

**Getting There**
The park is at the end of the Whangaparaoa Peninsula, 50km north of central Auckland by road. There is no public transport. The gates open at 6 a.m. every day and close at 9 p.m. in summer and 7 p.m. in winter. Access to the park is free. Take all your food and drink with you and remove all your rubbish.

**Habitats**
Pastureland, native forest, wetlands, beaches and sea cliffs.

**Best Time to Visit**
Summer (December, January and February) can be busy with beachgoers and campers; it is close to Auckland city. The birding is probably best in other seasons when the tracks are less busy.

**Useful Website**
www.aucklandnz.com/visit/discover/parks. At www.sossi.org.nz on the 'wildlife' tab, you can download excellent birders' information and a bird list.

This is a remarkable park in the diversity of ecological habitats that it has within it, given that it is not huge at only 500ha. Rolling grassland, dense rainforest, golden sandy beaches, dunes, swampy wetlands and ocean-edged cliffs provide a variety of habitats for numerous bird species. Seabirds and shags nest on the cliffs, rails, crakes and ducks enjoy the wetlands, passerines, ruru/Morepork and pigeons inhabit the rainforest with Paradise Shelducks, and there are New Zealand Pipits and pūkeko in the pastureland. Most of the park is protected from predation by a 1.7km predator-proof fence. Outside the fence, primarily beach, dunes and wetlands, the exterior of the park boundary is heavily trapped for predators.

## Tracks and walks

See pdf link to map: www.aucklandcouncil.govt.nz/parkmaps/map-shakespear-regional-park.pdf.

**Wetlands track** This starts at the Army Bay car park, and gives the best chance of seeing wetlands and shorebirds.

**Tiritiri Track** A 5km loop from Te Haruhi Bay car park, this track covers beach, pastureland and forest, and goes along the sea-facing clifftops. The best chance of seeing seabirds and nesting shags in cliff-edge trees is from the clifftops. There are magnificent views in all directions over the Hauraki Gulf, from numerous places. It is steep in parts and can be hot. Take plenty of water.

**Heritage Trail** This 4km trail takes in rainforest, a wetland and pastureland.

**Waterfall Gully** This 1km trail is a delightful rainforest stroll with an impressive waterfall near the end of it. It is good mostly for forest birds.

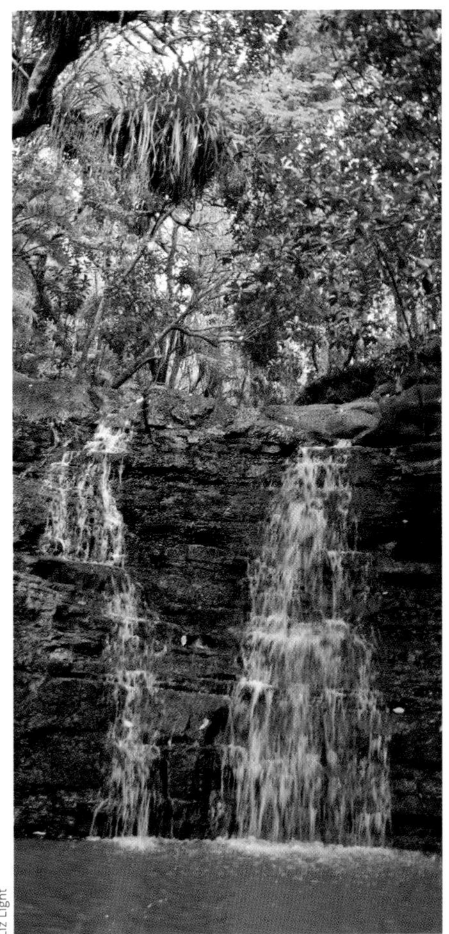

Liz Light

*The feature of the Waterfall Gully walk*

### Key Species

**Terrestrial:** korimako/Bellbird, Grey Warbler, pīwakawaka/New Zealand Fantail, kākāriki/Red-crowned Parakeet, Whitehead, North Island Robin, kererū/New Zealand Pigeon, tūī, New Zealand Pipit, Paradise Shelduck, pāteke/Brown Teal, Banded Rail, Australasian Shoveler, Grey Duck, Australasian Bittern, pūkeko, Sacred Kingfisher, Little Spotted Kiwi, Australasian Harrier, Shining Cuckoo, Long-tailed Cuckoo, ruru/Morepork, tauhou/Silvereye, tīeke/North Island Saddleback, Spotless Crake, Welcome Swallow.
**Shorebirds and seabirds:** Red-billed Gull, Black-backed Gull, Reef Heron, White-faced Heron, Variable Oystercatcher, Pied Oystercatcher, poaka/Pied Stilt, kororā/Little Penguin, Common Diving-petrel, Grey-faced Petrel, Pied Shag, Black Shag, Little Black Shag, Little Shag, Buller's Shearwater, Flesh-footed Shearwater, Fluttering Shearwater, Arctic Skua, Spur-winged Plover, Caspian Tern, White-fronted Tern, New Zealand Dotterel, Banded Dotterel, Bar-tailed Godwit. Numerous other shearwaters and petrels apart from those mentioned above can be seen at sea, in season.

Liz Light

*Pūkeko*

Gert op den Dries

## Kererū/New Zealand Pigeon

*Hemiphaga novaeseelandiae*

Kererū are big (650g), magnificent birds whose dark green and purple-coloured back and wings have a glossy metallic shimmer. The belly of the bird is pristine white, and it has a red bill and eyes. Kererū seldom vocalize, sometimes uttering a soft coo, but the swishing of their wings is distinctive. They do marvellous aerial acrobatics displays, seemingly at any time of the year, just for the fun of flying. Typically they fly very high, swoop almost directly downwards, then sweep out of the stall in an impressive arch before the ground. Kererū eat fruit, flowers and leaves, and are often seen in the forest areas of the park. Because they are large and eat whole fruit, they are an important vector in the spreading of seeds for the regrowth of native forest. They excrete seeds of up to 70 plant species, and have been known to get drunk on over-ripe taraire berries and fall out of their tree.

## Grey Warbler

*Gerygone igata*

The Grey Warbler, barely bigger than New Zealand's smallest bird, the Rifleman, miraculously thrives in a challenging environment (habitat loss and predation) and is common throughout and endemic to New Zealand. It lives in forest and forest edges, and regrowth mānuka is a favourite habitat. The species is grey-green above and off-white below with a pointed black bill and bright red eye – nothing special to look at but its song is a sweet trill. It hosts the parasitic Shining Cuckoo (see page 66), which lays its eggs in the warbler's nests. It is likely that warblers manage to raise a clutch of their own before the arrival of the cuckoos in spring, and therefore keep their species well populated.

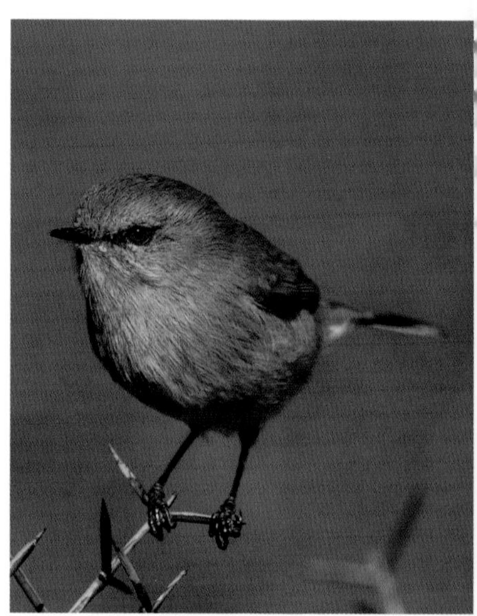

### Accommodation

Book a campsite at www.aucklandcouncil.govt.nz/parks-recreation/stay-at-park. Te Haruhi Bay Campground has all the usual camping ground facilities. Shakespear SCC Campground is for vehicles or caravans with self-contained toilet facilities. Both are beautiful beach-side locations. Besides great birding, there are three gorgeous beaches so you can swim, paddle board, or wind surf. It is wonderful to stay in the park at night to go kiwi spotting and hear the ruru call. There are numerous bed and breakfast accommodations nearby. See Airbnb or Booking.com.

# Muriwai Gannet Colony

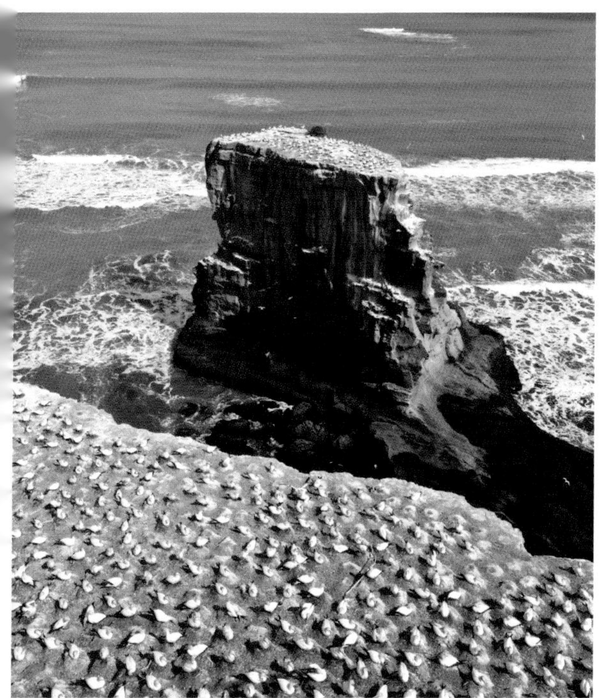

*Gannets populating the rock stack and nearby cliff*

## KEY FACTS

**Getting There**
Muriwai is 42km north-west of Auckland and is a 40-minute drive. There is no public transport. Weekends can be crowded as Muriwai is a famous surfing and swimming beach and lots of day-trippers come from the city.

**Facilities**
Sand Dunz Beach Cafe has excellent coffee and a good selection of food, including vegan and vegetarian.

**Best Time to Visit**
The gannets are here in August–April.
In August–September they are returning, finding mates with elaborate courtship rituals, squabbling over nesting spaces and gathering grass and seaweed to help construct the nests.
In November–December they are sitting on eggs.
In December–January the chicks are being fed.
In March–April the chicks are fledging. Then, by the end of April most of the birds have gone to their winter grounds in Australia.

The Muriwai gannet colony is designated an Important Bird Area (IBA) by BirdLife International. The gannetry is on an ocean-surrounded rock stack and two nearby clifftops at Otakamiro Point at the south end of magnificent Muriwai Beach. The gannetry, which includes hundreds of birds, is well fenced off but you can get to within a metre of some of the birds, which entirely ignore the humans behind the fence. The birds' elaborate courtship rituals include sparring with their beaks, interlinking their necks and mutual preening. They are fascinating to watch. The gannets are magnificent fliers but clumsy walkers. A bonus for birders is that between two parts of the gannetry there is a large and easily observable nesting site for hundreds of White-fronted Terns.

## Trails

The gannetry is on the cliffs at the south end of **Muriwai Beach**. It can be accessed by a path up the cliffs or from a side road, then a path along the tops of the cliffs and out to the Otakamiro promontary.

Liz light

White-fronted Tern

## Australasian Gannet
*Morus serrator*

This attractive bright white bird has an apricot head and black wing-ends and tail. It is common around New Zealand's coast and breeds in colonies on offshore islands and in isolated clifftop mainland areas. Gannets are big birds with wingspans of up to 2m and slender bodies 90cm long. They spend most of the winter foraging around New Zealand and southern Australian waters, and return to their colonies in August to mate, nest and rear their chick. They usually lay only one egg. Both parents incubate the egg and feed the chicks. Gannets are admired because of their beauty, elegant flying and audacious diving. They plummet, straight as an arrow, into the sea.

Liz Light

## Accommodation

At Muriwai Beach there is a range of accommodation, from rooms with shared bathrooms to luxury lodges. See Booking.com and Airbnb.

# Western Springs

*Western Springs is well known for waterbirds*

Western Springs park is 26ha of gardens, lake, wetlands, regenerating native forest and lawn just 4.5km from the heart of Auckland. To say it is a haven for waterbirds is an understatement; they reign supreme. The stale bread from people in the suburbs nearby ends up here, and swans, geese and ducks, in particular, seem to thrive on refined white flour. There are hundreds of them. The springs were the water supply for Auckland 120 years ago, and a large lake was constructed in the swampy land near the springs to act as the city reservoir. The city outgrew this water supply, and in 1961 the area was developed as a public park with careful planting of islands and wetlands in native vegetation. Now there is an around-lake walk. At the north end, more paths meander through the wetlands. Spring is gorgeous, with chicks of the numerous species that breed here being cute and easy to see. This is a busy park much loved and well used by people in the surrounding suburbs, so the birds are used to humans and not shy.

## KEY FACTS

**Getting There**
Western Springs is located at 859 Great North Road, 4.5km to the west of the city. Walk, cycle, or take a bus, car or taxi. This site is easy to get to. See Auckland transport at.govt.nz/bus-train-ferry.

**Facilities**
Toilets, barbecue sites, picnic tables, playgrounds, water fountains and fine walking paths.

**Habitats**
Lakes, swampy wetland, pastureland.

**Best Time to Visit**
Birds are here in all seasons, but late spring is sweet when the various species of chick have hatched.

## Tracks

The main **round-lake path** is 1.8km long, but there are additional loops and paths through wetlands and bush to the north of this. The paths are mostly paved and include bridges and boardwalks. The main track is wheelchair friendly. You cannot miss the waterbirds. They are everywhere. Because they are accustomed to humans, you can get quite close and bird photography can be excellent.

### Key Species

Pied Shag, Little Shag, New Zealand Scaup, New Zealand Dabchick, pūkeko, Australian Coot, Black Swan, Grey Duck, Paradise Shelduck, Red-billed Gull, Black-backed Gull. Besides these native species there are introduced domestic geese – hundreds of them – and Mallard ducks.

Liz Light

*Western Springs*

Oscar Thomas

*Paradise Shelducks, female (left), male (right)*

## Black Swan

*Cygnus atratus*

The Black Swan is gorgeously elegant on water and when flying but is a clumsy walker. It is a big bird, up to 140cm tall, and can have a 2m wingspan. It is found in groups in shallower lakes and rivers, including those in urban areas, where it may terrify children when it grabs food from them. Black Swans were abundant in New Zealand before human settlement but were hunted to extinction; they were an easy food source for Māori. They later self-introduced from Australia and are now abundant. There are plenty of them in Western Springs.

## New Zealand Dabchick

*Poliocephalus rufopectus*

The New Zealand Dabchick is endemic, tiny and shy, so is not often seen. It lives on small lakes, ponds and sheltered inlets of larger lakes in the North Island. The population is declining as its nests are susceptible to boat wash and predation by introduced mammals. Here, however, in Western Springs, they are thriving and are not so shy so are easy to see. They are pretty little birds, with dark blue-charcoal plumage and white eyes. The chicks have attractive black-and-white markings on the head and upper body, and when they are tiny they are commonly carried on the back of a parent.

Oscar Thomas

### Accommodation

Numerous options in Auckland city.

# Ambury Regional Park and Watercare Coastal Walkway

Liz Light

*Ambury Regional Park has superb birdwatching at high tide*

## KEY FACTS

**Getting There**
It is 20km from central Auckland by car and the journey takes half an hour. By public transport, from Auckland Central Railway station take the train to Onehunga and then the 313 bus to Mangere Bridge. There is then a pleasant foreshore walk to the park. Or from Wellesley Street, in central Auckland, take the 309 bus that stops at Mangere Bridge. Both journeys take about an hour. See Auckland Transport website planner: at.govt.nz/bus-train-ferry/journey-planner.

**Habitats**
Tidal estuaries, shell banks, scrubland, brackish lake, pastureland.

**Best Time to Visit**
High tide pushes the wading birds on to shell banks close to the hides so they are easy to see. In September migratory birds are returning to the area and in March they are gathering to depart. In winter species such as the Wrybill and tōrea/South Island Pied Oystercatcher make their way north from their breeding grounds in the South Island and overwinter in the harbour. There are resident birds all year round.

Ambury Regional Park, alongside the Manukau Harbour, in an area known as Mangere Bridge, is west of Auckland but part of the city. Ambury is a working farm, there are historical sites relating to early Māori activities in the area, and as it is on the flanks of a volcano there are interesting geographical features. Eighty-six bird species have been recorded at Ambury but is most prized for shorebirds and migratory waders. The 85ha park adjoins the equally large Watercare Coastal Walkway, along a knob of land that protrudes into the harbour. There are excellent walking and cycling tracks and numerous bird hides. Of the migratory Arctic waders that spend summer here, Eastern Bar-tailed Godwits, Lesser Knots and Ruddy Turnstones are most common, and can be seen in massive flocks at high tide. In winter, when they have migrated, New Zealand migrants arrive from the South Island – primarily poaka/Pied Stilts and Wrybills. There are resident colonies of Royal Spoonbills and Black Swans.

## Tracks and sites

See detailed park map: www.aucklandcouncil.govt.nz/parkmaps/map-ambury-regional-park.pdf.

**The foreshore walk** is 2km but there are side tracks with hides that are worth visiting.

**The Lost Gardens walk**, over grassland, is great for seeing pūkeko, some 50 of which are resident in the area, and poaka/Pied Stilts, in the winter season, as well as the remnants of traditional Māori gardens.

**Ambury Regional Park** joins seamlessly with Watercare Coastal Walkway, and here, on side paths off a loop track and along the harbour edge, is a great place to see flocks of the migratory birds and resident shorebirds, as well as Royal Spoonbills and Black Swans. March is especially superb for birding as thousands of Asian migratory birds gather here before their long flight north. Eastern Bar-tailed Godwits, Lesser Knots and Ruddy Turnstones are the most common.

**Pūkaki Lagoon** is a circular explosion crater surrounded by a tuff ring that, besides being a pristine volcanic feature, is tidal, and poaka/Pied Stilts, oystercatchers, Grey-faced Herons and Black Swans can be seen here. It is 600m across and takes half an hour to walk around.

### Key Species

Australasian Gannet, Black Shag, Pied Shag, Little Shag, Little Black Shag, Black Swan, Paradise Duck, Grey Duck, Grey Teal, Australasian Shoveler, White-faced Heron, Reef Heron, kōtuku/White Heron, Little Egret, Royal Spoonbill, Australasian Harrier, Pied Oystercatcher, pūkeko, Variable Oystercatcher, tōrea/South Island Pied Oystercatcher, New Zealand Dotterel, Banded Dotterel, Black-fronted Dotterel, Spur-winged Plover, Wrybill, Ruddy Turnstone, Lesser Knot, Curlew Sandpiper, Sharp-tailed Sandpiper, Red-necked Stint, Eastern Curlew, Whimbrel, Eastern Bar-tailed Godwit, Black-tailed Godwit, Arctic Skua, Southern Black-backed Gull, Red-billed Gull, Caspian Tern, White-fronted Tern, poaka/Pied Stilt, Welcome Swallow, tauhou/Silvereye, New Zealand Pipit, kererū/New Zealand Pigeon, pīwakawaka/New Zealand Fantail, Grey Warbler, tūī.

*South Island Pied Oystercatchers*

Oscar Thomas

## Bar-tailed Godwit

*Limosa lapponica*

Bar-tailed Godwits nest in Alaska but spend the summer months in New Zealand harbours. Manukau Harbour hosts a significant number of them. They are large waders (up to 600g), with dappled brown feathers on top and creamy-grey below. They have long legs and long, slender bills. Bar-tailed Godwits are often seen at high tide, particularly before departure, in huge flocks. On their way to Alaska they have a feeding stop-off around the Yellow Sea. On their return to New Zealand, in September, they fly 11,000km from Alaska to the Manukau Harbour. They take 8–10 days to do this, without sleep or feeding, at an average speed of 56km an hour. They are then at half their weight and have been flying for so long that they are unable to fold their wings. For a few days, they stroll around foraging with their wings limp, open by their sides.

There are approximately 75,000 Bar-tailed Godwits but the numbers are dropping. The problem is habitat pressure in the feeding stop-off on the journey around the Yellow Sea. Massive reclamation projects, pollution and overfishing are some of the issues.

## Pūkeko

*Porphyrio melanotus*

Pūkeko are conspicuous birds, as big as a domestic hen, with long red legs, a red beak and purple-blue breast plumage. They live in groups near fresh water and in nearby grassy fields. The group is usually dominated by a lead female, and they often nest cooperatively with up to three hens laying eggs in the same nest. Various members of the group care for the chicks. The chicks are cute: black fluff-balls on long, strutting legs. Pūkeko are aggressive towards predators and other birds, are not endangered and are widely distributed. They are native but not indigenous, and are also found in parts of Australia and in some Pacific islands.

### Accommodation

There is a fine flat, sheltered camping area. Book a campsite at www.aucklandcouncil.govt.nz/parks-recreation/stay-at-park. There are numerous bed and breakfast accommodations nearby. See Airbnb or Booking.com.

# Motutapu Island

*Motutapu Restoration Trust has done massive replanting on the island*

Motutapu Island, a 1,510ha island in the Hauraki Gulf, 25 minutes' ferry ride from Auckland, has excellent birding, beautiful walks and heavenly beaches for swimming and picnicking. It was well populated by Māori for over 100 years, until Rangitoto, an adjoining volcanic island, exploded 600 years ago and covered most of the island in volcanic ash. Post-European settlement it was farmed extensively and most of the forest that remained after the volcanic eruption was burned or milled. DOC took over the island in the 1990s and the process of eliminating pests and replanting native vegetation began. It is still farmed (by DOC), but in a sustainable way, and hundreds of hectares are fenced off from grazing to protect the native vegetation, beaches and waterways from damage by farm animals. It was declared predator free in 2011.

The Motutapu Restoration Trust works with DOC to restore and replant native ecosystems, including wetlands and forest that will create a safe haven for native plants and animals. The work, all done by volunteers, ranges from collecting eco-sourced seeds, removing weeds and planting trees (over half a million so far), to reintroducing species. There has been translocation of many endangered birds, and other species, such as korimako/Bellbirds and kākāriki, have returned naturally.

## KEY FACTS

### Getting There

There is a ferry service to Home Bay on weekends and public holidays, and daily ferry departures to Rangitoto Island, from where you can walk around the coast and cross the causeway to Motutapu. See Fuller Ferry timetables at www.fullers.co.nz/timetables-and-fares. On the island you will be rewarded with an abundance of easy walking tracks, beaches, scenic lookouts and great birding.

### Facilities

There are toilets at Home Bay.

### Habitats

Forest for passerines and other forest birds, pastureland for pipits, pūkeko and more, beaches for shorebirds and estuarine wetlands for Banded Rails, and Reef and White-faced Herons. Seabirds nest on the cliffs.

### Best Time to Visit

Summer holidays (January and February) can be busy with beach-goers and campers from Auckland. In July and August pick your weather as it can be cold and wintry. Many of the birds are resident, so all other seasons are good.

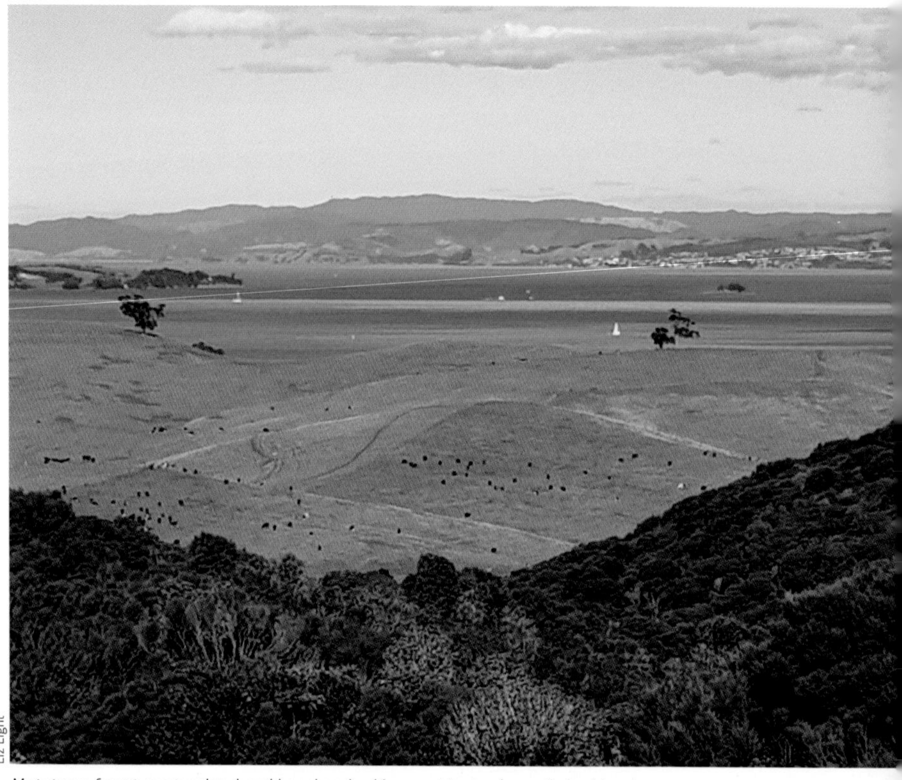


*Motutapu; forest, pastureland and beaches, looking west towards south Auckland*

The island has a variety of habitats – forest for passerines and other forest birds, pastureland for pipits, pūkeko and more, beaches for the likes of Shore Plovers, dotterels and oystercatchers, and estuarine wetlands for Banded Rails, and Reef and White-faced Herons. Seabirds nest on the cliffs.

## Tracks and sites

There are many to choose from so it depends on what kinds of birds you are interested in.

**Motutapu Walkway** The walk from the causeway to Rangitoto and Home Bay (Motutapu ferry arrival point) takes 1½ hours on an easy track past areas of plant restoration where a variety of birds is present. Watch for Saddleback and kākāriki.

**Wetland Track** This covers wetland/pastureland birds, including pūkeko and takahē.

**Beaches** (Sandy Bay, Waikarapupu Bay). Look for dotterels, Shore Plovers, Black-backed Gulls, terns of various kinds and pipits.

**Emu Point Track, Waikarapupu Bay Access Track** and **Billy Goat Point Track, Pōhutukawa Track.** On these walks, from high points above the rocky shore, look out for Reef Herons and seabirds.

**Home Bay** If you have access to a boat or a kayak Home Bay has superb birding. Sea and shore birds can be seen on the beach and wharf, takahē and pūkeko stroll on the grass and the bush behind is full of passerines singing like there is no tomorrow. Hīhī, tūī, bellbirds, saddleback, pīwakawaka/ New Zealand Fantail, Silvereye and Grey Warblers all add to the chorus.

There are numerous freshwater ponds with Spotless Crake and pāteke/Brown Teal.

New Zealand Pipit

## Key Species

**Terrestrial:** Little Spotted Kiwi, korimako/Bellbird, Grey Warbler, pīwakawaka/New Zealand Fantail, takahē, Whitehead, tīeke/Saddleback, kererū/New Zealand Pigeon, Paradise Shelduck, tūī, Tomtit, Spotless Crake, pāteke, Shining Cuckoo, Sacred Kingfisher, ruru/Morepork, pūkeko, Welcome Swallow, Australasian Gannet.
**Shorebirds and seabirds:** Red-billed Gull, Reef Heron, White-faced Heron, Reef Heron, Variable Oystercatcher, kororā/Little Penguin, Pied Shag, Little Shag, Little Black Shag, Black Shag, Spotted Shag, Buller's Shearwater, Flesh-footed Shearwater, Fluttering Shearwater, Caspian Tern, White-fronted Tern, New Zealand Dotterell, Shore Plover, Spur-winged Plover, Arctic Skua, Southern Black-backed Gull.

## Reef Heron

*Egretta sacra sacra*

This blue-grey, medium-sized heron is better known in South Asia and the Pacific, but there is a stable population of some 500 or more birds in the north of New Zealand. They have a heavier bill and shorter legs than most herons, and fish around shoreline rocks and rock pools. They nest in a loose collection of cupped twigs often high on a rock stack, or in a rocky crevice.

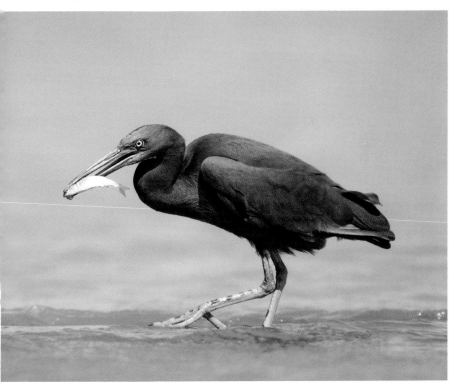

## Shore Plover

*Thinornis novaeseelandiae*

Motutapu Island is the easiest place to see these severely endangered shorebirds, and they can be found on some of the more isolated beaches. They are colourful little birds with an orange-red beak, black face-mask, white breast and underparts, and grey-brown crown and upper body. They weigh about 60g and have relatively short legs for shorebirds. There are only 250 of these birds, with 17 resident on the island. Successful nesting here increases the population each year. Before mammalian predators arrived they lived around most of New Zealand's shores, but by 1960 there was only one population on a small predator-free island in the Chathams group. Captive breeding and release have halted the population decline, but they remain susceptible to avian predation. At their peril, they tend to wander from their safe island home and are sometimes seen on very unsafe Auckland beaches.

Oscar Thomas

## Accommodation

The Home Bay campsite is divided into five different areas called pods, offering a total of 38 non-powered tent sites. No fires are allowed at any time. There is no power so you will need to bring a torch. There are no showers, but water is available. The cost is minimal but bookings are essential. Visit DOC's campground site, www.doc.govt.nz/parks-and-recreation/places-to-go/auckland/places/motutapu-island-recreation-reserve/things-to-do/home-bay-motutapu-island-campsite.

# Pūkorokoro Miranda Shorebird Centre

Shorebirds have a safe haven at the Pūkorokoro Miranda Shorebird Centre

Pūkorokoro Miranda Shorebird Centre, on the Firth of Thames, is a terrific shorebird area, particularly at high tide when the birds gather near the shore. The reason for this is the area's geographical uniqueness. An active shell and sand chenier plain, with numerous shell ridges with mudflats between, some 15km long, stretches from the coastline 2km into the Firth of Thames. In this vast habitat of fecund mudflats and roosting shell banks, shorebirds thrive. In the Firth there are 85,000ha of mudflats. Behind this, along the coast, there is low-lying grassland, brackish swampland and freshwater ponds that provide a fine habitat for the likes of egrets, spoonbills, bitterns and crakes. Arctic migratory shorebirds arrive here in their thousands in spring, the Red-necked Stint being the smallest and the Bar-tailed Godwit the most common. In the winter months migratory birds from the South Island such as Pied Oystercatchers and the Wrybill live here.

Three endemic species breed here; the New Zealand Dotterel, Variable Oystercatcher and Black-billed Gull. Other breeding species include the poaka/Pied Stilt, Spur-winged Plover, White-faced Heron, Banded Rail, Black-backed Gull, White-fronted Tern and pūkeko.

## KEY FACTS

**Getting There**
The centre is on the Firth of Thames 60km south of Auckland. There is no public transport and it takes an hour to get there by car. Drive south along the East Coast Road.

**Facilities**
The Shorebird Centre is an information and education facility open every day, and offering up-to-date birding information seven days a week. You can make tea and coffee here but food is not available.

**Habitat**
Shell banks, tidal mudflats, muddy estuary, swampland, scrubland.

**Best Time to Visit**
There is good birding all year round, but the most birds are seen in January–March. The Arctic migratory birds return in September and leave in March. Two hours before or after high tide is the best time to see the birds. There is a link to the daily tide times on the website.

**Useful Websites**
Pūkorokoro Miranda Shorebird Centre has a great website with most things that visitors need to know and replies promptly to emails, see www.miranda-shorebird.org.nz admin@shorebirds.org.nz.

Liz Light

*Pied Stilt*

**Key Species**

Buller's Shearwater, Fluttering Shearwater, Australasian Gannet, Black Shag, Pied Shag, Little Shag, Little Black Shag, Spotted Shag, Black Swan, Paradise Duck, Grey Duck, Grey Teal, Australasian Shoveler, White-faced Heron, Reef Heron, kōtuku/White Heron, Cattle Egret, Royal Spoonbill, Australasian Bittern, Australasian Harrier, Banded Rail, Spotless Crake, Pied Oystercatcher, pūkeko, Variable Oystercatcher, New Zealand Dotterel, Banded Dotterel, Large Sand Dotterel, Black-fronted Dotterel, Spur-winged Plover, Golden Plover, Wrybill, Ruddy Turnstone, Knot, Sanderling, Curlew Sandpiper, Sharp-tailed Sandpiper, Red-necked Stint, Eastern Curlew, Whimbrel, Little Whimbrel, Eastern Bar-tailed Godwit, Black-tailed Godwit, Terek Sandpiper, Arctic Skua, Southern Black Backed Gull, Red-billed Gull, Caspian Tern, White-fronted Tern, poaka/ Pied Stilt.

## Trails and hides

The website has an excellent map. www.miranda-shorebird.org.nz/wp-content/uploads/2017/01/Miranda-Map-2017.jpg.

The walk from the **Shorebird Centre** to the hides at the south end of the reserve is just over 2km but there is good viewing of wetland birds along the way. There are three hides towards the south end with spectacular viewing over mudflats and banks, and if the tide is high you can see thousands of birds feeding either ahead of the incoming tide or roosting when it is high You are not able to wander randomly as, for the sanctity of the birds, there are restricted areas. Respect this and stay on the paths.

Liz Light

*Kōtuku*

Oscar Thomas

## Wrybill

*Anarhynchus frontalis*

The Wrybill, a small, quirky endemic plover, is the world's only bird whose beak turns laterally; it bends to the right. This makes the white-bellied, grey-backed Wrybill look odd, as if it bumped into something at speed and bent its beak. In fact, this is a useful adaptation to allow the bird to wedge its bill under round river stones to extract insects hiding there. Wrybill breed in the braided rivers of Canterbury. After breeding, in January–July, they migrate to the north, primarily to the Firth of Thames and the Manukau Harbour. There are approximately 5,300 Wrybill, 40 per cent of which overwinter near Miranda. The population is slowly declining. Most threats occur during the nesting seaon and they include predation, changing river systems through water depletion because of irrigation and four-wheel driving on the gravel beds.

## Red Knot/Lesser Knot

*Calidris canutus*

This medium-sized sandpiper is a migrant, breeding in east Siberia and returning in September to the northern harbours and Farewell Spit for the New Zealand summer. Most of the year it has white undersides and grey-patterned top sides with a black beak and legs. In February and March, while still in New Zealand, it moults into a rich reddish breeding plumage, hence the name. It is a flocking bird and tends to feed in groups on the tidal flats, and the birds roost together. Red Knots are the second most prolific migratory bird in the Miranda area, after Bar-tailed Godwits, and there are about 30,000 in New Zealand. The population is declining primarily because of habitat degradation in Bohai Bay, in the Yellow Sea, on the East Asian–Australasian Flyway.

## Accommodation

The Shorebird Centre has overnight accommodation; there are self-contained units as well as bunk rooms for hire, all at very reasonable rates. See the website. Miranda Holiday Park is nearby, offering cabins, camping and motorhomes facilities. Moreover, it has natural mineral hot springs. There is accommodation in cottages and rooms nearby. See booking.com and Airbnb.

# Sanctuary Mountain, Maungatautari

Liz Light

*Sanctuary Mountain*

## KEY FACTS

**Getting There**
Manu Tioriori, Sanctuary Mountain Maungatautari visitor centre, is at 99 (the top of ) Tari Road, Pukeatua, Waikato.

**Facilities**
There are toilet facilities, informative displays and information about guided tours. You can also buy refreshments and souvenirs. Open from 8.30 a.m. to 4 p.m. daily. Closed 25 December.

**Fees**
The main mountain is free. See the website for the cost of the Southern Enclosure and the guided tours of the Tautari wetlands.

**Habitat**
Mature conifer/broadleaf/podocarp forest, some wetland, adjoining pastureland.

**Best Time to Visit**
All seasons are good as this site does not have many migratory birds. Spring usually has more bird activity.

**Useful Websites**
www.sanctuarymountain.co.nz; info@sanctuarymountain.co.nz.

The Sanctuary Mountain project began with a dream to protect the diversity of plant and animal species living on Maungatautari Mountain. The community, including landowners, local iwi and local residents, came together to restore and protect this precious ecosystem, and in 2001 the Maungatautari Ecological Island Trust was formed. Sanctuary Mountain comprises 3,400ha and the resulting 47km-long pest-proof fence that encircles it is the largest in the world. This extraordinary fence offers a safe haven for many of New Zealand's most endangered species.

Most of the mountain is mature conifer/broadleaf/podocarp forest and is now an ecosystem as close to the pre-human New Zealand environment as possible. Due to the size and quality of the habitat it is a viable home for self-sustaining populations of many of New Zealand's most endangered animals, from birds to bats, frogs to reptiles, and giant wētā. To quote the website, 'Step through the pest-proof gate into another world where ancient New Zealand forest towers majestically above well-maintained walking tracks and the air is filled with birdsong. Enjoy a variety of hikes and bush walks around our eco sanctuary, take a guided tour or nature walk and learn about why this conservation project is so special.'

## Tracks and trails

**The Southern Enclosure** is accessible during Manu Tioriori visitor centre opening hours. There are numerous tracks and trails and it is easy to enjoy six hours exploring them all. You will be given a map when you buy your ticket. Within the southern enclosure there is a 16m viewing platform taking you high in the canopy. You can view the understorey, sub-canopy and canopy of the ancient forest flora and the birds residing in it.

**Tautari wetlands** can only be visited on a guided tour.

**The Wairere Traverse** (Over the Mountain) track and the **Northern Enclosure** are accessible at any time of the day, but for the safety of visitors and of the flora and fauna, visiting after dark is not recommended. Follow this link for in-depth information and to download a PDF of a map: www.sanctuarymountain.co.nz/over-the-mountain.

**Hike up the mountain** in the mid-afternoon (three hours) to a viewing point. Enjoy the view, the sunset and the evening bird chorus. Note the call of kōkako and korimako/Bellbirds, in particular, and walk back down the same trail to the visitor centre. Take a good torch with you.

Oscar Thomas

*Welcome Swallow*

### Key Species

Australasian Harrier, Australasian Shoveler, Banded Rail, tūī, korimako/Bellbird, Black Shag, Spotted Shag, Little Shag, New Zealand Dabchick, pīwakawaka/New Zealand Fantail, Grey Duck, Grey Teal, Grey Warbler, hīhī/Stitchbird, kārearea/New Zealand Falcon, kererū/New Zealand Pigeon, Sacred Kingfisher, Long-tailed Cuckoo, ruru/Morepork, North Island kākā, Northern Brown Kiwi, kōkako, Paradise Shelduck, New Zealand Pipit, pūkeko, North Island Robin, tīeke/North Island Saddleback, Shining Cuckoo, tauhou/Silvereye, Spotless Crake, Tomtit, tūī, takahē, Welcome Swallow, Whitehead, White-faced Heron.

*Sanctuary Mountain Maungatautari, a vast predator-proof sanctuary for birds*

## North Island Robin

*Petroica longipes*

The North Island Robin has sooty upperparts, a creamy breast, long, slender legs and an upright stance. It is about the size of a sparrow. Its song is a busy, many-noted *cheep* with an occasional trill, and it sings to define territory and attract mates. The birds forage on the forest floor and tree trunks for vertebrates, but eat fruit and seeds in season. They are not shy of people and will hop about on bush tracks at the feet of walkers. This is a common bird to see and hear while in the sanctuary.

Gert op den Dries

## Tīeke/North Island Saddleback

*Philesturnus rufusater*

These distinctive birds, the size of blackbirds, have dark bodies with an orange saddle over the back and similarly coloured wattles that enlarge as they age. The species is endemic and has been particularly prone to predation by introduced mammals because it nests and roosts in cavities in trees and rocks. Tīeke are a conservation success story. By the 1960s there was only one remaining population, of some 500 or so birds, on an island in the north Hauraki Gulf. Since then they have been successfully translocated (or have naturally colonized) to 20 predator-free locations and are now thriving, with many thousands of birds. They eat invertebrates, nectar and fruit, and are enthusiastic foragers with strong beaks with which they pick into rotten wood and bark, and strong legs for foraging in leaf litter. While foraging they are often accompanied by pīwakawaka/New Zealand Fantails and Whiteheads, which enjoy the tiny insects their feeding activities disturb.

## Accommodation

Out in the Styx (styx.co.nz) is in the country next to the mountain. It has a bunk house and 10 rooms with en suite bathrooms, as well as great home-grown food. There is also plenty of accommodation in Cambridge, Tirau and other villages nearby. See booking.com and Airbnb.

# Sulphur Bay Wildlife Refuge, Rotorua

Sulphur Bay Wildlife Refuge occupies 145ha of much larger, 80km² Lake Rotorua. This area of the lake is geothermal with such features as steaming pools, boiling mud and large, flattish pans of hard sulphurous rock. It is easy to access as it is less than 1km to the centre of Rotorua, a major tourist city. Sulphur Bay Wildlife Refuge is unique in the world in that it is a geothermal area and many birds, both visiting species and residents, are attracted to the bay and adapted to it because of the geothermal warmth. This helps them to conserve the energy that would usually be needed to keep their bodies warm. Two species of gull nest on the sulphur flats, and the warmth from the earth helps keep the eggs and chicks warm. The refuge is home to many resident bird species, including the nationally threatened New Zealand Dabchick, Banded Dotterel and Black-billed Gull. Some 60 species can be seen here.

The water in the bay is low in oxygen and quite acidic. Birds seldom feed in the bay as few insects and larvae can tolerate the acidic conditions. It provides a safe and warm place to nest and roost, and is close to food sources in other fecund freshwater parts of the lake. Red-billed and Black-billed Gulls are present in large numbers, and both nest on the sulphur flats. Their presence here is unusual as it is some distance from the sea. Three species of shag roost and nest on islets and trees near the shore and fish in the main body of the lake. Black Swans, New Zealand Dabchicks and New Zealand Scaup tend to gather in the convergence zone where the acidic water meets fresh water. They thrive here and are present in large numbers.

## KEY FACTS

**Getting There**
Sulphur Bay is less than 1km from the centre of Rotorua. Get a map at the Rotorua I-site Visitor Information Centre, 1167 Fenton St, Rotorua. It is a pleasant walk or cycle from here through Government Gardens to Sulphur Bay at the south end of Lake Rotorua.

**Facilities**
There is a toilet block at the boat-launching ramp car park at the north end of the sanctuary. There is a cafe at Polynesian Spa, not far from the walking path along the lake edge.

**Habitat**
Sulphuric volcanic flats, rocky pans, acidic and fresh water convergence zone, freshwater lake.

**Best Time to Visit**
Spring and early summer to see the gulls nesting in large numbers, but any time is good.

Liz Light

*Red-billed Gull nesting colony*

Liz Light

*Red-billed Gull nesting colony*

Liz Light

*Black-billed Gull nesting colony*

## Tracks and sites

There is an excellent track around this part of the lake. The boat-launching ramp area at the north end of the bay, known as **Sulphur Point**, has (among other species) a large population of Black Swans, New Zealand Dabchicks, New Zealand Scaup and shags (two or three species), which usually roost on the nearby trees and rocks. This is a convergence zone between the acid thermal water and fresh lake water, and it seems to be a great food source for diving waterbirds.

From here you can walk on a **well-formed path/boardwalk** on or near the lake edge in a southerly direction closer into the heart of the bay. The path winds its way between various forms of geothermal activity. It is dangerous to deviate because you can accidently step into a steaming fumarole or boiling mud. Tūī and introduced passerines inhabit some of the bush between the lake and the path. Red-billed Gulls nest on rocky shore areas, Black-backed Gulls can be seen on islets and there is a shag roost on an islet not for from the shore.

Pass Polynesian Spa and follow the path south. This is the gull nesting and roosting area. There are hundreds of birds, some very near the path. There are Black-billed Gulls in the Sulphur Bay car park area and their nesting colony is on the flats, across a thermal stream, not far away.

### Key Species

Black Swan, Paradise Shelduck, New Zealand Scaup, New Zealand Dabchick, Grey Teal, Australasian Shoveler, poaka/Pied Stilt, Little Shag, Little Black Shag, Pied Shag, White-faced Heron, pūkeko, Banded Dotterel, Spur-winged Plover, Southern Black-backed Gull, Red-billed Gull, Black-billed Gull, Caspian Tern, Sacred Kingfisher, Grey Warbler, pīwakawaka/New Zealand Fantail, tūī, tauhou/Silvereye, Welcome Swallow, New Zealand Pipit.

*Black Swan and Scaup*

Liz Light

## Black-billed Gull
*Chroicocephalus bulleri*

The Black-billed Gull is endemic and superficially looks like its red-billed cousin but is quite different. Though there are still many thousands of Black-billed Gulls, the population is declining because it is threatened in many ways, but primarily from land-use changes and predation. The species is a colonial nester, usually in the gravel riverbeds of the South Island, and unlike its red-billed cousin, it seldom begs or is found in urban areas. The large colony in Sulphur Bay is atypical in that it is close to the centre of Rotorua, a major city, and nowhere near a braided river or the sea. This is a great place to observe these attractive birds. They are all around the Sulphur Bay car park, are quite used to humans, and their territory, along with that of the Red-billed Gulls, is encroaching on the walkway. They nest in large numbers on the Sulphur Flats nearby, taking advantage of the warm earth. Black-billed Gulls do not tend to scavenge but they can be seen feeding on insects and worms on agricultural land in the Rotorua area.

## Little Black Shag
*Phalacrocorax sulcirostris*

Little Black Shags are black, sleek and glossy, with dark grey bills and black legs and feet. They are similar in size to Little Shags, but lack any white plumage and have a comparatively long tail. Both species can be seen roosting in the same trees near the boat-launching ramp, at the northern end of Sulphur Bay. The birds are found in harbours, lakes, estuaries and coastal inlets, and feed on small fish in fresh water. They are gregarious when feeding and roosting, and often forage cooperatively in flocks. They are found predominantly in the North Island, but the population is increasing and they are spreading in the South Island.

## Accommodation

Rotorua is a tourist hotspot and there are all sorts of accommodation options within a few kilometres, from expensive hotels through to camping grounds and backpackers' lodges.

# Ohiwa Harbour

Vast mudflats provide rich pickings for shorebirds

Ohiwa Harbour, in the eastern Bay of Plenty, covers 26km$^2$. It is protected by two sandspits, Ohope Spit (11km) to the west, and Ohiwa Spit (1km) on the east side. Within it, and on the beaches on the ocean side of the spits, there is a variety of habitats for birds. You can spend a good deal of time getting to know this harbour and its birds. Bird numbers in the harbour range from about 1,200 in winter to 5,200 in summer. Eighty species have been recorded here, including 36 Arctic waders. Of these, a dozen species are annual visitors, the most prolific being the Bar-tailed Godwit. Up to 5,000 of these birds arrive each spring and 200–300 overwinter.

Nukuhou Wetlands

## KEY FACTS

### Getting There

Ohiwa Harbour is in the eastern Bay of Plenty. Whakatane and Ohope are the nearest towns in the west and Opotiki in the east. There is a bus service from numerous towns to Whakatane. See www.intercity.co.nz. You will need transport when you get there. There are plenty of car-hire places in Whakatane. Go birdwatching via self-guided walks and kayaking and driving around harbour margins. Kayaks can be hired at Ohope Beach and Ohiwa Family Holiday Park, and are available from some accommodation places.

### Top spots

Ohiwa Spit at high tide, around the harbour margins at low tide, Nukuhou Saltmarsh at dawn or dusk, and Tern Island, from a distance, from the end of Ohiwa Spit or Ohope Spit.

### Facilities

Whakatane has all the facilities of a major town, including cafes, restaurants, banks, ATMs, shops, photography specialists, supermarkets, rental vehicle and taxi companies.

### Best Time to Visit

September–April for the international migratory birds. An annual Birds-a-Plenty Festival to welcome the migrants is held in early October. Late summer to August for New Zealand migrants. Numerous bird species are resident so birdwatching is good all year round.

There are numerous islands in the harbour. Tern Island Wildlife Refuge is the most important for birds. It is a breeding area for the New Zealand Dotterel, White-fronted Tern, Variable Oystercatcher, White-faced Heron, Fernbird, Caspian Tern, and Red-billed and Black-backed Gulls. Other birds seen in season are the Reef Heron, Bar-tailed Godwit, Asiatic Whimbrel, Eastern Curlew, Little Tern, Spur-winged Plover, poaka/Pied Stilt, and Black, Pied and Little Black Shags. On the beaches on the ocean side of both spits look for Red-billed and Black-billed Gulls and Black-backed Gulls. White-fronted Terns, Caspian Terns and Skua can often be seen. Gannets nest on White Island, on the horizon, so they can be seen diving for fish. There are four types of shag in the harbour and they can often be seen around the Ohiwa Wharf. Fernbirds, Banded Rails, Spotless Crakes, bitterns, weka and pūkeko are found in the reeds and saltmarshes in the inner harbour. The area around the Nukuhou Wetlands has been intensely trapped for predators since 2005 and these secretive saltmarsh dwellers are doing well here.

## Tracks and sites

**Ohiwa Spit** The area is a high-tide roost for godwits, along with other migrants, and an important nesting site for the New Zealand Dotterel and Variable Oystercatcher. The spinfex and tussock is also home to Fernbird and Banded Rail. At low tide, scan the mudflats for various migrant species feeding among the flocks of godwits. At high tide the birds roost on Ohiwa Spit, the adjacent Ohiwa Beach and Tern Island. Take care not to disturb nesting or roosting birds, particularly in spring and summer. Stay on tracks and, on the beach, remain below the high-water mark. Meg and Mike Collins,

Liz Light

*Ohiwa Harbour has a variety of habitats*

Caspian Tern

of Ohiwa Spit Care Group, live nearby and are happy to guide and chat about the local birdlife. Email: mcollins658@gmail.com.

**Nukuhou Marshlands** This area has a lovely little saltmarsh walk, and in the mornings and evenings there is a high chance of seeing or hearing the Fernbird, Banded Rail, Australasian Bittern, Spotless Crake, Marsh Crake and weka. Other native birds seen here are the Australasian Harrier, korimako/Bellbird, tūī, Grey Warbler, ruru/Morepork, pīwakawaka/New Zealand Fantail, kererū/New Zealand Pigeon, tauhou/Silvereye, Sacred Kingfisher, New Zealand Pipit and Shining Cuckoo. Estuary birds seen include the Black Shag, Little Black Shag, Pied Shag, Reef Heron, White-fronted Heron, Caspian Tern, Pied Silt, kōtuku/White Heron and pūkeko. Since trapping for predators began in 2005, the Fernbird population has gone from about 10 birds to more than 100. Stuart Slade, of Nukuhou Saltmarsh Care Group, lives nearby and is happy to guide and chat about the local birdlife. Email: s.m.slade@xtra.co.nz.

**Tern Island Wildlife Refuge** This tiny 11ha island is a wildlife refuge that supports a significant number of bird species. There is no public access because of its importance as a bird breeding and roosting site, but there is good viewing by kayak/boat an hour on either side of high tide. Birds likely to be seen include the Bar-tailed Godwit, Caspian Tern, White-fronted Tern, Black-billed Gull, Variable Oystercatcher, New Zealand Dotterel and Red-billed Gull. Pied Shags roost at the western end and a small group of about 20 Fernbirds nests in the undergrowth.

### Key Species

Bar-tailed Godwit, Lesser Knot, tōrea/South Island Pied Oystercatcher, New Zealand Dotterel, Banded Dotterel, Fernbird, poaka/Pied Stilt, Variable Oystercatcher, Pacific Golden Plover, Turnstone, White-fronted Tern, Black-billed Gull, Red Billed Gull, Wrybill, Reef Heron, Spotless Crake, Bittern, Banded Rail, pūkeko, Paradise Shelduck, Caspian Tern, Spur-winged Plover, White-faced Heron, weka, tūī, kererū/New Zealand Pigeon, Spotless Crake, Marsh Crake, Shining Cuckoo, Black Shag, Little Black Shag, Pied Shag, Reef Heron, pūkeko, ruru/Morepork.

## Fernbird

*Bowdleria punctata*

The Fernbird is the only species in its family in New Zealand. It is endemic and widely spread throughout wetlands, including reedbeds, saltmarshes, swamps, and damp scrub and damp forest. Numbers have decreased because of the draining of wetlands for pasture and predation by introduced mammals. The birds are poor flyers and prefer to scramble. Fernbirds do not exactly sing, but utter a raspy repeated *tick* with, sometimes, an intermittent squeak. They are sparrow sized (24g), and have creamy underparts and chestnut upper bodies with chocolate dark spots and semi stripes. They are attractive and shy and blend well into their habitat. The birds can be heard, and seen by those who persist, at the Nukuhou Marshlands and Ohiwa Spit.

Oscar Thomas

Oscar Thomas

## Weka

*Gallirallis australis*

This endemic, flightless rail is about as big as a hen but more streamlined. It has red eyes, a pointed tail and strong, short legs, and is mottled brown though the colours vary between the four different subspecies. The North Island subspecies, after years of decline, is making a population recovery because of trapping of predators and more awareness. It is vulnerable to predation by mustelids and dogs. They are friendly, cheeky birds, but have not helped themselves by messing up vegetable gardens and stealing domestic hens' eggs. They often roost in pairs at night and duet, uttering a sweet plaintive call that reminds me of my childhood near Gisborne. They are increasing in numbers in and near the Nukuhou Marshlands.

## Southern Black-backed Gull

*Larus dominicanus*

This large black-backed gull is found on most coastlines in New Zealand and does not mind nesting on top of the high-rise buildings in Auckland city. It is plentiful, handsome and will scavenge, hunt and pester other seabirds for their food. It is also common around ports and landfills. It is heavy bodied, with a black back, white underparts and big yellow bill. It shows pretty black and white marking when the wings are folded, and a pointy tail. The juvenile is mottled brown, and the parents continue to regurgitate food for it when it is as big as them and can fly. The juveniles look completely different from the parents and are often mistaken as a different species. The gulls are found on the sea sides of both Ohope and Ohiwa Spits, and inside as well.

## Accommodation

There are a couple of harbour-side cottages available for rent. www.fantailcottage.co.nz near the Ohiwa Spit is great for birders. It has big harbour views and easy access to the Ohiwa Spit, and the hosts are keen and knowledgeable birders. Ohiwa Family Holiday Park also offers different accommodation options (ww.ohiwaholidays.co.nz). On the Ohope Spit, between the ocean beach and the harbour, Top Ten Holiday Park (www.ohopebeach.co.nz) offers campsites, motel units, apartments and cabins. There are many accommodation options in Whakatane, Ohope and Opotiki, including motels, facilities for backpackers, cottages and lodges, and camping grounds.

# Whirinaki Te Pua-a-Tāne Conservation Park

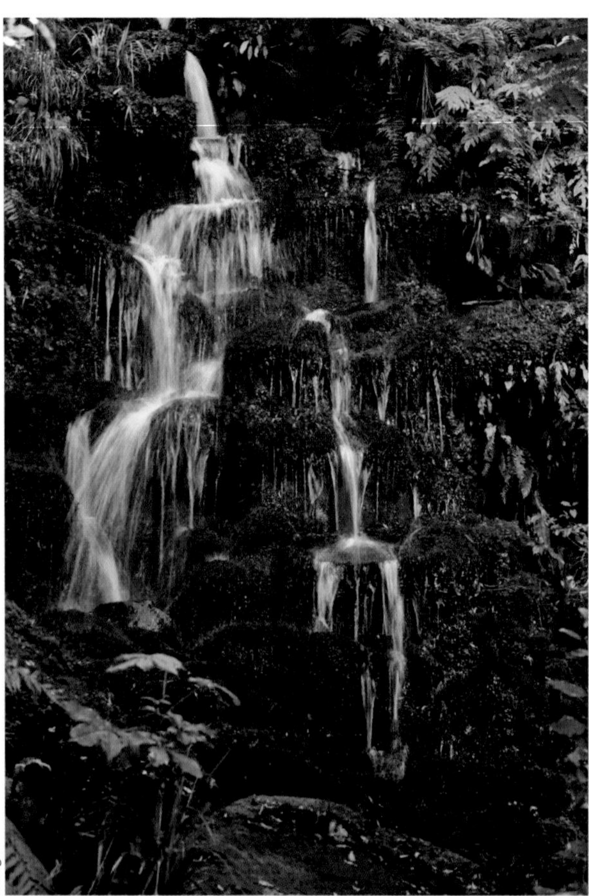
Liz Light

*One of numerous waterfalls in this primeval forest*

## KEY FACTS

**Getting There**
Whirinaki Te Pua-a-Tāne Conservation Park is 90km south-east of Rotorua, off Te Whaiti Rd, via SH38. The most popular entrance to the park is past Minginui village and up River Road to the River Road car park, the starting point for many walks.

**Weather**
The weather can be unpredictable and visitors should be prepared for cold, wet spells even in summer. Frosts and snow are also likely on high ground throughout the year. High rainfall causes rivers to rise quickly, although they usually fall rapidly once the rain stops. You need to be well prepared for tramping and walking in the park.

**Facilities**
There are toilets at the River Road car park and well-maintained tracks.

**Best Time to Visit**
There is good birding in all seasons but more bird activity and song in spring. Winter can be cold and stormy.

Majestic, primeval, ancient, awe-inspiring – Whirinaki Te Pua-a-Tāne Conservation Park has the finest of New Zealand's podocarp rainforests and is of international significance for its biodiversity. It is 60,000ha in extent, the size of Singapore. Podocarps are trees that have linear or scale-like leaves and are usually dioecious (having male and female flowers on separate plants of the same species). There are eight native species in New Zealand and the best known, the five forest giants, are dioecious. In Whirinaki the podocarp forest can be lush with a dense undergrowth f shrubs, ferns and tree ferns. There is lso monoao and mānuka land on the frost ats. Herb fields, grassland and shrubland re present on riverbeds and in forest earings, and subalpine shrubland exists n the high ridges and peaks. The altitude is 60–1,365m. There are beautiful river flats nd rolling, tree-covered hills and gullies, s well as steep, rugged country. Besides npressive forest, there are gorgeous fast-owing rivers and streams, tranquil lagoons nd stunning waterfalls.

Vhirinaki is an ecological hotspot with huge diversity of plant, insect and bird pecies, and ecologists, botanists and bird anciers from around the world revere this lace. The birdlife is increasingly impressive ach year and there is intensive trapping of redators and the occasional poison-based erial eradication programme that targets he mustelids and rats that wreak havoc on ests. The Tomtit, North Island Robin, kākā, Vhitehead, Shining Cuckoo, Long-tailed Cuckoo, kākāriki/Yellow-crowned Parakeet, vhio/Blue Duck, Spotless Crake, New Zealand Dabchick and Little Black Shag can usually be seen by amateur birders, and the Northern Brown Kiwi is often heard at night.

Liz Light

*Podocarp crowns form a dense canopy*

*Spotless Crake*

## Tracks and walks

This is a big park and there are many tracks.

**Arohaki Lagoon Track** four hours return, offers New Zealand Dabchicks, Spotless Crakes and other waterbirds. You can camp here.

**Waiatiu Falls Track** 1.5 hours return, follows the river. You can see whio/Blue Ducks, and a host of forest birds.

**Whirinaki Waterfall Loop** 4 hours return. You can see whio/Blue Ducks, and numerous forest birds. Gorgeous forest walk.

There are 3–5-day walks with accommodation in huts. Visit the Rotorua I-site or the above website.

The area is huge and local birders know where to find individual species. Both these companies can tailor-make guided birdwatching walks in Whirinaki: Foris Eco-tours, www.foris.co.nz; Whirinaki Experiences, whirinaki.com.

### Key Species

Northern Brown Kiwi, Paradise Shelduck, kākāriki/Red and Yellow-crowned Parakeets, kākā, whio/Blue Duck, Grey Duck, New Zealand Scaup, New Zealand Dabchick, Little Black Shag, Little Shag, Black Shag, White-faced Heron, Australasian Harrier, kārearea/New Zealand Falcon, Tomtit, kererū/New Zealand Pigeon, ruru/Morepork, tūī, Sacred Kingfisher, North Island Robin, kākā, Whitehead, Shining Cuckoo, Long-tailed Cuckoo, Spotless Crake, Grey Warbler, pīwakawaka/New Zealand Fantail, Welcome Swallow.

Liz Light

*Grey Teal*

Gert op den Dries

## Whio/Blue Duck

*Hymenolaimus malacorhynchos*

The whio is one of only three of the world's species of torrent-living ducks and is rarely found far from fast-flowing streams. It mates for life and the ducklings are fed by both parents. Its numbers plummeted due to predation by introduced animals, especially stoats, and through competition for food with introduced trout. However, the Whio Forever project, a combination of community volunteers, business sponsorship and DOC know-how, has stopped the decline, and the bird's numbers are increasing. This was achieved though intensive predator trapping and other forms of habitat management. In Whirinaki the comeback has been particularly impressive, and there are now more than 50 pairs on 17km of the Whirinaki River and more in side-streams and waterways.

## Kākāriki/Red-crowned Parakeet

*Cyanoramphus novaezelandiae*

## Yellow-crowned Parakeet

*C. auriceps*

These two endemic emerald-green parakeets look similar but are separate species. The Red-crowned kākāriki is bigger, more raucous, favours open land on forest edges, and will forage on the ground. The Yellow-crowned kākāriki (see also page 130) is more common but shyer and less likely to be seen. It favours the canopy in tall, unbroken forest. Both species feed on berries, seeds, fruit and insects, and nest in holes in trees. Both are thriving in Whirinaki since intensive trapping removed their main predators. Stoats can take out whole nests, including the female when she is sitting on a nest, and rats predate eggs and chicks.

## Accommodation

Whirinaki Recreation Camp has 3 cabins each of which sleeps up to 30 people. There are 3 different camping grounds with vehicle access each with toilets. Within the park there are 9 tramping huts with access by foot only. Each type of accommodation has a different pricing structure and each should be booked. See the Department of Conservation website below. It is possible to book online or by phone. www.doc.govt.nz/parks-and-recreation/places-to-go/east-coast/places/whirinaki-te-pua-a-tane-conservation-park/

# Pureora Forest Park

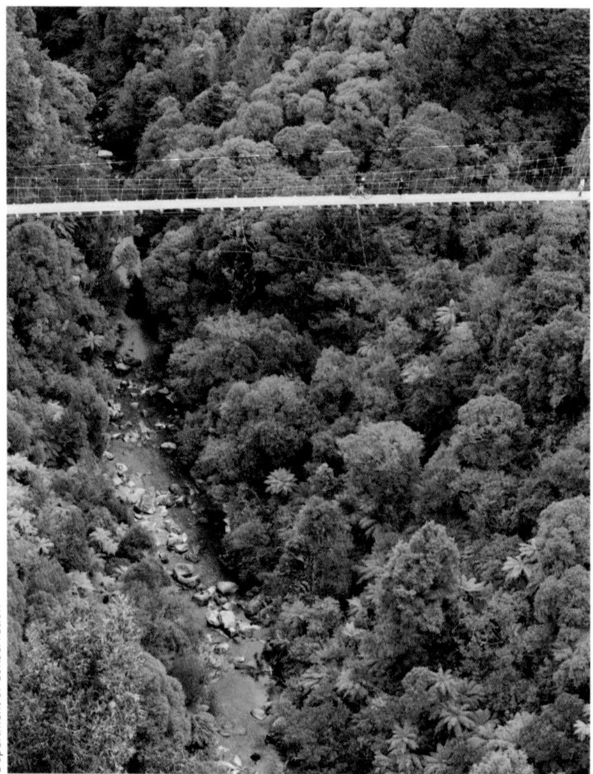

Department of Conservation

*A swing bridge for walkers and cyclists over forest and stream*

This forest park is huge, covering 760km$^2$, and is one of the finest rainforests in the world. It contains the largest remnant of giant podocarp forest in New Zealand. Due to the variety in terrain, the effect of volcanism and altitudinal range of 400–1,200m, there is a wide range of forest types and habitats, including wetlands, rivers and, at higher altitudes, scrubland. Pureora was the site of a major conservation struggle in 1978, with conservationists trying to save the forest against loggers attempting to make a living from felling it. The conservationists won, eventually, and now Pureora village is the main base for access to the park, and old logging tracks are now walking and cycling tracks. There is a good variety of forest and wetland birds in the area, and it is especially rewarding to spot them in this pristine Jurassic environment. There are many great places for birding but the most accessible are near Pureora village.

## KEY FACTS

**Getting There**
Pureora Forest Park is remote but not difficult to get to. There is no public transport and it is best to stay in the area for at least one night to access the park for the dawn chorus and birdwatching. It lies between Te Kuiti, Taumarunui and Lake Taupo in the central North Island, and is easily accessed by State Highway 30 and State Highway 32.

**Facilities**
There is no retail shop or petrol station at Pureora; the nearest facilities are at Benneydale, 15 minutes west of Pureora along SH 30. Cafes are thin on the ground but shops in Benneydale sell the basics.

**Habitat**
Wide variety of forest types and habitats, including wetlands, rivers and, at higher altitudes, scrubland.

**Best Time to Visit**
There is good birding in all seasons but more bird activity and song in spring. Winter can be cold, stormy and muddy.

**Useful Websites**
Google Pureora Forest Park and you will find lots of information about this place.

## Tracks and walks

There are about 20 tracks in this vast forest. For a link to a map with many of them, see www.doc.govt.nz/globalassets/documents/parks-and-recreation/tracks-and-walks/waikato/pureora-forest-park-brochure.pdf.

**Waipapa Walk**, a loop walk of about one hour gives you the best chance of seeing a variety of birds. This links to excellent information about that walk: www.doc.govt.nz/globalassets/documents/parks-and-recreation/tracks-and-walks/waikato/waipapa-loop-walk.pdf. As dawn is the best time for birding, this involves staying in Pureora or nearby. The dawn chorus can be amazing, with kōkako, korimako/Bellbirds and tūī all adding their part to the melody. In the scrubland at the beginning of the walk listen for Fernbirds; they are shy but they are there. Keep an eye out for the New Zealand Pipit. Next, the path weaves through marvellous stands of kahikatea forest and gorgeous giant tree ferns. Deeper into the podocarp forest watch out for the elusive kōkako and the more common tūī, korimako/Bellbird and pīwakawaka/New Zealand Fantail. Further on, where there is a good covering of leaf litter on the ground, you can see North Island Robins, Tomtits and Whiteheads. In the open area by the Waipapa Stream you will often see flocks of kākā flying high and screeching, and kārearea/New Zealand Falcons are often seen here.

### Key Species

Northern Brown Kiwi, Paradise Shelduck, whio/Blue Duck, Grey Duck, kārearea/New Zealand Falcon, Australasian Harrier, Spur-winged Plover, pūkeko, Spotless Crake, kererū/New Zealand Pigeon, kākā, kākāriki/Red-crowned Parakeet, Long-tailed Cuckoo, Shining Cuckoo, ruru/Morepork, Sacred Kingfisher, Grey Warbler, kōkako, Tomtit, North Island Robin, Welcome Swallow, korimako/Bellbird, tūī, pīwakawaka/New Zealand Fantail, New Zealand Pipit, Whitehead, Rifleman, Fernbird, tauhou/Silvereye.

*Tomtit*

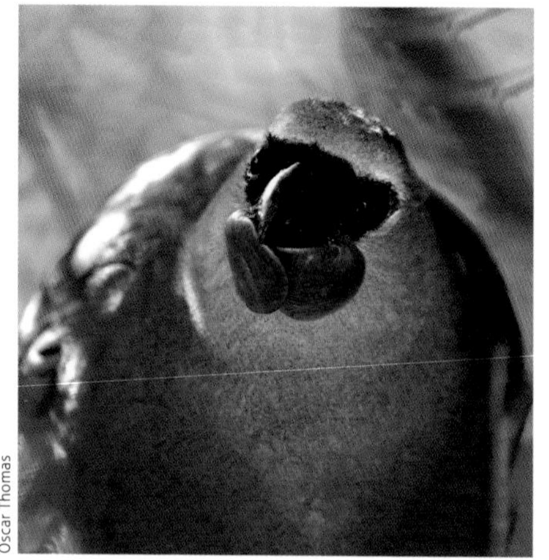
Oscar Thomas

## North Island Kōkako

*Callaeas wilsoni*

The North Island Kōkako, sometimes known as the Blue-wattled Crow, is endemic and rare, living in 24 North Island locations. It is medium sized, and blue-grey with bright blue wattles, a black mask, and black beak and legs. It is a shy bird, referred to as the grey ghost, but famous for its exquisite song. When the birds duet the sound is strong and pure, stretching over many notes, like the ring of a piano tuner's fork or the clear, fine notes of an organ. They often sit high in the canopy and sing, though they forage in under layers. They sing during mating and to defend territory, and require 5–25ha of territory. The species persists because of pest control: ship rats and possums eat their eggs and chicks. Pureora Forest has one of the largest populations, with approximately 100 pairs. They are considered to be at risk but recovering.

## New Zealand Pipit

*Anthus novaeseelandiae*

This slender, brown-patterned, medium-sized (35g) songbird is easily identified because of its erect stance when still and the constant flicking of its tail when it forages. It is a ground forager and is bigger and more upright-standing than sparrows, but has a similar colouring. Pipits enjoy open habitats such as scrub, pasture and coastlines. This species is an unassuming but delightful and alert bird that is relatively unafraid of humans.

### Accommodation

This is a remote area and the park is vast but there are a few accommodation options near or in Pureora village. Pureora Cabins are the closet to the Waipapa walk. Check on the internet to see what else is available in the area. It is important to stay the night so you can go birding in the morning, at first light. There is camping at various places in the park and there are DOC huts to stay in. See the website.

# Lake Rotokare Scenic Reserve

*Morning glory*

## KEY FACTS

**Getting There**
The reserve is 12km east of Eltham, South Taranaki. Turn off at Eltham on to King Edward Street and follow this road, which becomes Rawhitiroa Road. Sangster Road is on the right, with good signs for the reserve.

**Facilities**
Toilets, information centre and picnic tables. The nearest shop and petrol station are in Eltham.

**Habitat**
Swamp forest, lake, wetlands and marginal vegetation.

**Best Time to Visit**
There is good birding in all seasons but more bird activity and song in spring.

**Useful Website**
www.rotokare.org.nz.

Lake Rotokare Scenic Reserve is a stunning 230ha area of forested hill country with extensive wetlands and an 18ha lake. The reserve is surrounded by a predator-proof fence, creating a pest-free sanctuary with diverse habitats for birds, including the lake, wetlands, swamp forest and bush. Mature tawa, rewarewa and māhoe dominate the forest, which is home to the tūī, korimako/Bellbird, kererū/New Zealand Pigeon, Grey Warbler and North Island Robin, plus a variety of other bird species. The lake-edge habitat consists of raupo, flax, and pukatea and kahikatea swamp forest, and is home to notable birds such as the Fernbird and Spotless Crake.

## Tracks

A **4km lake walkway** circles the lake. The first 600m is boardwalk. The walkway makes its way through magnificent lowland forest before heading deep into Rotokare's swamp forest. Besides a variety of birds, you will see the swamp-forest giant trees, kahikatea and pukatea, with gnarly buttress roots. In winter areas of the track can be wet and muddy. There is also a wetland boardwalk and floating viewing platform for water and wetland bird viewing.

*Rotokare walkway*

### Key Species

Australasian Shoveler, korimako/Bellbird, Northern Brown Kiwi, White-faced Heron, New Zealand Scaup, Fernbird, Shining Cuckoo, Australasian Harrier, Black Swan, kārearea/New Zealand Falcon, Grey Warbler, Sacred Kingfisher, kereū/New Zealand Pigeon, poaka/Pied Stilt, Welcome Swallow, Whitehead, ruru/Morepork, hīhī/Stitchbird, tūī, North Island Robin, North Island Tomtit, Black Shag, Little Shag, tīeke/North Island Saddleback, New Zealand Dabchick, pūkeko, Spotless Crake, pīwakawaka/New Zealand Fantail, tauhou/Silvereye.

Gert op den Dries

### Ruru/Morepork
*Ninox novaeseelandiae*

This is the only species of native owl, and it has four subspecies. It is dark brown on top with chestnut and brown dapples on the underside. Ruru have big yellow owl-eyes and a small hooked bill and, when roosting, are rugby-ball shaped. Some say that the birds' two-note hoot sounds like *ruu ruu*, while others swear that it is *more pork* – either way, it is the sound of the night that means home to New Zealanders. The owls are blessedly abundant and there is nothing quite as delightful as the parents giving their young flying lessons, with three just-fledged chicks lined up nervously on a branch and plucking up the courage to spread their wings and fly.

### Kārearea/New Zealand Falcon
*Falco novaeseelandae*

The New Zealand Falcon is New Zealand's fastest flier and can exceed 100km per hour when after prey. It usually lives in forest or mountain pasture and tussocks, and hunts live prey such as small birds, but will go for rabbits, young hares and ducklings. It is now a protected species and the population is recovering. It is adaptable and has been introduced to vineyard areas where it acts as a natural way to keep small birds from eating ripening grapes.

Gert op den Dries

## Accommodation

No camping in the reserve. The nearest accommodation is in Eltham, though there might be some homestays closer. The city of New Plymouth is 50km away and there are many other small, pretty country towns with a variety of accommodation near Eltham.

# Bushy Park Sanctuary, Whanganui

Bushy Park Sanctuary

*Walking track through Bushy Park*

**KEY FACTS**

**Getting There**
The sanctuary is at 791 Rangitatau East Road, 8km from Kai Iwi and 25km north of Whanganui city. Access to the forest is during daylight hours only.

**Facilities**
Public toilets and well-formed tracks. The Homestead is open for Devonshire teas, functions, events and accommodation, Tuesday–Sunday, plus public holidays, at 10 a.m. to 4 p.m.

**Habitat**
Virgin lowland forest.

**Best Time to Visit**
There is good birding in all seasons but more bird activity and song in spring.

**Contact**
www.bushyparksanctuary.org.nz.

Bushy Park Sanctuary is a 100ha predator-free native bird sanctuary, set among a patch of virgin lowland forest near Whanganui. A wide variety of birdlife thrives in the sanctuary. You will be able to see or hear species such as korimako/Bellbirds, kererū/New Zealand Pigeon, North Island Robins, tīeke/North Island Saddlebacks, hīhī and ruru/Morepork, as well as kārearea/New Zealand Falcons, pīwakawaka/New Zealand Fantails, Grey Warblers, pūkeko, tauhou/Silvereyes, Sacred Kingfishers and White-faced Herons. The sanctuary is home to kiwi too.

ellbird

### Key Species

Northern Brown Kiwi, Paradise Shelduck, Little Shag, White-faced Heron, Australasian Swamp Harrier, kārearea/New Zealand Falcon, pūkeko, Southern Black-backed Gull, kererū/New Zealand Pigeon, kākā, Shining Cuckoo, Long-tailed Cuckoo, ruru/Morepork, Sacred Kingfisher, tīeke/North Island Saddleback, hīhī/Stitchbird, Grey Warbler, Whitehead, korimako/Bellbird, tūī, pīwakawaka/New Zealand Fantail, North Island Tomtit, North Island Robin, Fernbird, tauhou/Silvereye, Welcome Swallow, New Zealand Pipit.

ree species include māhoe, mamaku, ukatea, rātā and rimu, along with colonies f ferns and mosses. A feature of the reserve a large northern rātā named Rātānui ('Big ātā'). It is estimated to be 500–1,000 years ld, it is 43m in height and has a girth that xceeds 11m.

s well as the sanctuary, the property eatures a large heritage Edwardian homestead. In 1962 G. F. Moore bequeathed Bushy Park to the Royal New Zealand Forest & Bird Protection Society and hence the public of New Zealand.

## Tracks

There are 3.4km of well-formed walking tracks throughout the forest, providing easy all-weather access.

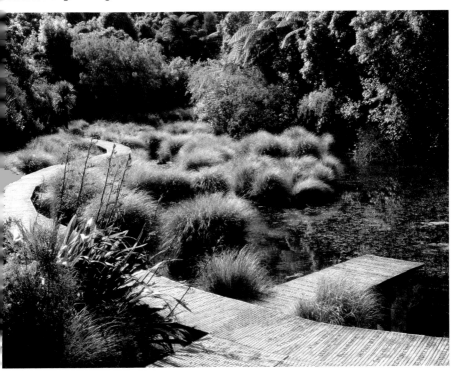

*All-weather access paths at Bushy Park*

Glenda Peake

## Sacred Kingfisher
*Todiramphus sanctus*

The Sacred Kingfisher is the only species in this family that is native to New Zealand, and it is abundant. It is often seen on power lines, high trees and fence tops above estuaries, ponds, rivers and streams, beady-eyeing whatever potentially tasty treat might move in the water. When it flies it is a fast-moving flash of buff and turquoise. It nests in burrows in clay banks, cliff faces and knots of trees. The birds are strident defenders of their nests and will dive-bomb cats, rats and other birds that stray too near.

## Pīwakawaka/New Zealand Fantail
*Rhipidura fuliginosa*

The pīwakawaka is endemic, much-loved and abundant. It is tiny with a proportionately very large fantail, is not shy of people and will accompany them on walks, flitting hither and thither in seemingly erratic ways while softly, incessantly tweeting. Despite predation, it has fared well since European settlement as it can adapt to native forest, introduced commercial pine forest, scrubland, hedges and even suburban gardens. It does not do well in severely cold or wet winters. Its aerial acrobatics are fascinating to watch – the birds are smile-makers.

### Accommodation

Homestead luxury accommodation. With six rooms to choose from, tastefully decorated, each room is unique and stylish. Numerous hotel-style amenities but no cafe. For family holiday accommodation, the bunkhouse offers families/groups self-contained shared accommodation that can sleep up to 11 people. Cook in the full kitchen. For campers and motorhomes, powered sites and non-powered sites are available, with toilet and cooking facilities provided. Tel.: 06 342 9879, email: bushyparkbbwanganui@gmail.com.

# Manawatū Estuary

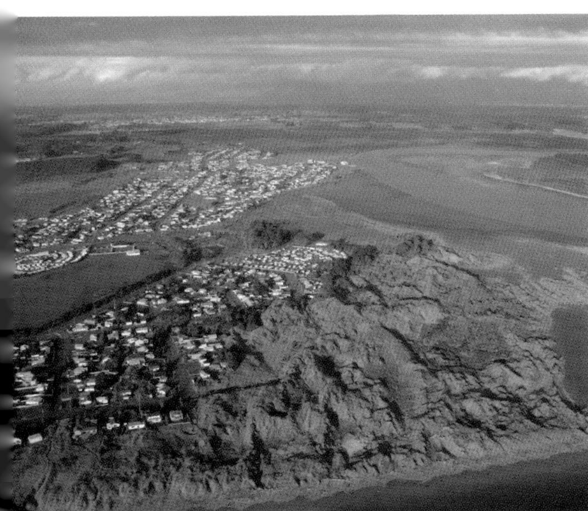

he Manawatū Estuary next to the town of Foxton

## KEY FACTS

**Getting There**
The Manawatū Estuary lies next to the small town of Foxton Beach, 5km from Foxton on State Highway 1. From Seabury Ave, in central Foxton Beach, go down Dawick Street on the left. This leads to a viewing platform, a great spot from which to see waders roosting at high tide. Otherwise, you can access the estuary by walking along the beach towards the river mouth. Kayaking at high tide also provides a great way to see birds. Keep an eye on the tides to make sure you do not get stranded. Just because an area is accessible at low tide, this does not mean that you will enjoy getting back at high tide.

**Habitat**
Beach, sand dunes, mudflats, salt marshes and salt meadows.

**Best Time to Visit**
October to March to see the migratory wading birds.

**Useful Website**
The estuary has an excellent website so refer to this first: www.massey.ac.nz.

'he Manawatū Estuary lies at the mouth of the Manawatū ːiver on the west coast of the lower North Island. It was .esignated a Wetland of International Importance in July 005, and is one of six wetland areas in New Zealand to have ;ained RAMSAR status. The site comprises about 200ha and ncludes regions of sand dunes, mudflats, salt marshes and alt meadows. It is home to 13 species of threatened bird, and he mudflats are a feeding ground for many Arctic migrants, vhich make an annual trip from Alaska and Siberia to New Zealand for the summer. A number of other migratory birds ıre occasionally seen at the estuary. One hundred and ten bird species have been recorded here.

*The estuary provides varied habitats of sand dunes, mud flats, salt marshes and salt meadows*

## Tracks and sites

See this link on the website for two excellent maps: www.massey.ac.nz/~grapson/metrust/ontheestuary.php. The viewing platform on **Dawick St** is a great place from which to see waders close up, roosting at high tide. Seabirds and shorebirds can be seen from the beaches on both sides of the river mouth. Kayaking is a good way to see the birds. See map for boat-launching ramp.

Liz Light

*White-faced Heron*

### Key Species

Though 110 bird species have been recoded here, the following can often be seen.
**Waders:** Wrybill, Bar-tailed Godwit, Red Knot, Banded Dotterel, Pacific Golden Plover, poaka/Pied Stilt, Cattle Egret, Pied Oystercatcher, Variable Oystercatcher.
**Wetland birds:** White-faced Heron, Royal Spoonbill, Fernbird, Black Swan, Grey Duck, Grey Teal, New Zealand Shoveler, Sacred Kingfisher, pūkeko.
**Seabirds:** Black-backed Gull, Red-billed Gull, Black-billed Gull, Caspian Tern, White-fronted Tern.

## Caspian Tern

*Hydroprogne caspia*

The Caspian Tern may be seen on the beach, on the mudflats or flying over the water searching for fish. It is a large, stocky bird with a black cap that touches the bill in breeding plumage, and a slight crest. In non-breeding plumage the black cap may be flecked with white. The legs and feet are black, while the thick beak is bright red. The upperparts of the body are grey and the underparts are white. When feeding the Caspian Tern flies 5–10m above the water, and on spotting a fish it dives in.

Liz Light

## attle Egret
*rdea ibis*

nis Australian migrant arrives in New ealand in autumn to spend the winter ere. It is a small but stocky egret, which hen standing has a hunched posture. It is oout the same size as a White-faced Heron, ut not as slim and lacks the long neck. In non-breeding plumage, which is how it occurs in New Zealand, it is pure white. It has grey legs and a straight yellow bill, with the yellow extending to the yellow facial skin. Cattle Egrets are common in Asia, arrived in Australia in 1948 and were first seen in New Zealand in 1963. They can be found over much of lowland New Zealand and flocks are loyal to particular areas.

Oscar Thomas

## Accommodation

There is plenty of holiday-home accommodation around on the likes of booking.com and Airbnb. There are motels at both Foxton and Foxton Beach.

# Ahuriri Estuary, Napier

Glenda Peake

*Pied Stilt*

## KEY FACTS

**Getting There**
Napier is a city on the east coast of the North Island. You can fly there, take a bus or drive. The estuary is on the edge of the city, adjacent to the airport and main highway, and is large and unmissable.

**Habitat**
Wide range of estuarine habitats, from fresh to salty, shallow to deep, and sandy to muddy.

**Best Time to Visit**
October to March to see the migratory wading birds.

**Useful Website**
There is good information on the DOC website: www.doc.govt.nz/parks-and-recreation/places-to-go/hawkes-bay/places/napier-area/ahuriri-estuary/ahuriri-estuary-bird. Or visit DOC, 59 Marine Parade, Napier, Tel.: 06-834 3111. You can obtain a range of maps and several topographical sheets that include diagrams of walking tracks.

Ahuriri Estuary, comprising 470ha on the edge of the city of Napier, offers a wide range of estuarine habitats, from fresh to salty, shallow to deep, and sandy to muddy, to support more than 70 bird species, including shorebirds, migratory waders, waterbirds and others. It is a fecund feeding ground for birds, with more than 20 fish species and an array of whelks and molluscs. Some species overwinter to take advantage of the estuary's abundant food supplies, and migratory birds rest and recover here after their exhausting journey from the Arctic. The area is protected, along with Westshore Wildlife Reserve, which is separated from the estuary by a state highway. Napier city is well aware of the value of this estuary and closely monitors the quality of storm water that drains into it. Like other good birding sites adjacent to cities, it has the problem of human interference and cats and dogs; it also has the advantage of a large group of keen conservationists and birders who look out for it, as their place.

## racks and sites

his huge area, though all linked around the tuary, is divided into a number of different eas. The estuary includes adjacent areas ıch as Southern Marsh, Lower Channel, Vestshore Lagoon, Northern Pond and andcorp Marsh. Each provides a different abitat and has different bird species, epending on the season and the tides, and ach is accessible by road, to some extent, ıen by walking or cycling. There are many alks and birdwatching sites in these areas, oo many to list in this book. The best way to ccess information and maps relevant to the umerous tracks is by visiting DOC, Napier.

Oscar Thomas

*Little Black Shag*

### Key Species

Black Shag, Pied Shag, Little Shag, Little Black Shag, Australasian Gannet, Paradise Duck, Grey Duck, Australasian Shoveler, White-faced Heron, Australasian Bittern, pūkeko, Spur-winged Plover, Pied Oystercatcher, Black-billed Gull, White-fronted Tern, Caspian Tern, poaka/ Pied Stilt, Sacred Kingfisher, Eastern Bar-tailed Godwit, Golden Plover, Red-necked Stint, Curlew Sandpiper, Grey-tailed Tattler, Sharp-tailed Sandpiper, Royal Spoonbill, kōtuku/ White Heron, Wrybill.

*Bar-tailed Godwits*

Geoff Moon

## Australasian Bittern

*Botaurus poiciloptilus*

The Australasian Bittern is found in large areas of swampland but is reclusive so is not easy to see. It is more likely to be heard than seen; the male makes a loud booming sound in the breeding season. The population is declining due to habitat loss (the draining of swampland), predation of chicks and eggs by introduced mammals, and decreasing water quality and reduced food availability. The birds live in tall, dense reedbeds and freshwater wetlands, and feed mainly on fish, along with freshwater crayfish, lizards, frogs and insects. In the estuary they may be heard and sometimes seen around reedbeds.

## Poaka/Pied Stilt

*Himantopus leucocephalus*

Pied Stilts are long-legged, slender, black-and-white wading birds, which self-introduced from Australia in the early 1800s and have thrived. There are now an estimated 30,000 living in wetlands including swamps, estuaries, saltmarshes, rivers and lakes. They are aggressive toward predators, especially during nesting, hence the large and stable population. They feed on invertebrates: insects and worms when on land, and aquatic insects and larvae when in ponds, swamps and estuaries. Ahuriri Estuary provides a variety of habitats that suits them well, and the population here is in the hundreds. Pied Stilts are sociable and tend to stay in groups, often with other waders, feeding and roosting in large, noisy flocks. They nest on shingle banks, riverbeds, sand dunes and pasture.

### Accommodation

This wetland is on the doorstep of a decent-sized city and beautiful tourist area famous for wineries, seafood and fresh farm produce. There are hundreds of accommodation options, including backpacker facilities, bed and breakfasts, motels, hotels and exclusive lodges.

# Boundary Stream and Shine Falls

Shine Falls

Boundary Stream Mainland Island Reserve is an excellent birdwatching site and has some stunning scenic walks, especially if you include Shine Falls, which, at 58m, is the highest waterfall in the province. DOC manages Boundary Stream as a mainland island, and with the help of many volunteers there is intensive predator control.

Birdlife is abundant, including the North Island Robin, Northern Brown Kiwi and kōkako, with more recent releases of kākāriki/Yellow-crowned Parakeets and kākā. Other birds that can be seen include the kārearea/New Zealand Falcon, Tomtit, Whitehead and kererū/New Zealand Pigeon. Grey Warblers and Riflemen are common, and in late spring, when they have arrived from their winter home in the South Pacific, both Long-tailed and Shining Cuckoos are often seen or heard.

## KEY FACTS

Getting There

1. **Shine Falls** This is a 1½-hour drive from Napier. From Napier head east on SH2 and follow this to Tutira, then turn left into Matahorua Road. After 11km turn left into Heays Access Road, which has a gravel surface. Follow it for 6.5km until you reach a DOC car park.
2. **Boundary Stream Mainland Island** Like for Shine Falls, follow State Highway 2 to Tutira, turn left at Tutira on to Matahorua Road, then left on to Pohokura Road.

Facilities

Picnic tables, a shelter and a toilet at the DOC car park.

Best Time to Visit

There is good birding in all seasons but more bird activity and song in spring. Tracks can be muddy in winter.

Useful contact

DOC, 59 Marine Parade, Napier. There is a range of maps and several topographical sheets that include diagrams of walking tracks. Tel.: 06 834 3111.

## Tracks

This information is taken from the DOC website, www.doc.govt.nz/parks-and-recreation/places-to-go/hawkes-bay/places/boundary-stream-area.

**Tumanako Loop Track** An interpretation nature walk; 40 minutes return. This gentle loop track is great for children. It offers a variety of forest types and among the more common forest birds to be sighted are kākā near the car park. Listen for kōkako singing.

**Heays Access Road to Shine Falls Track** 1½ hours return. This popular track, suitable for children, wanders past castle-like limestone formations and into deep forest before emerging at the base of the stunning Shine Falls. The habitat includes mixed lowland forest with a variety of bird species within it.

**Kāmahi Loop Track** Two hours return. There are large areas of kāmahi forest and some podocarp forest at the beginning of the track. Along the bluff tops, aerial displays from kererū/New Zealand Pigeon, tūī and korimako/Bellbirds are often visible.

**Bell Rock Loop Track** Three hours return. This track features a variety of vegetation and seasonal bird activities. The track climbs through mixed beech and podocarp to a forest, and the low canopy provides opportunities to view large numbers of tūī, korimako/Bellbirds and kererū/New Zealar Pigeons.

### Key Species

Northern Brown Kiwi, Australasian Harrier, kererū/New Zealand Pigeon, Rifleman, kākā, kōkako, Long-tailed Cuckoo, Shining Cuckoo, ruru/Morepork, Sacred Kingfisher, Grey Warbler, tīeke/North Island Saddleback, kōkako, Tomtit, North Island Robin, korimako/Bellbird, tūī, pīwakawaka/New Zealand Fantail, Whitehead, tauhou/Silvereye, kākāriki/Yellow-crowned Parakeet.

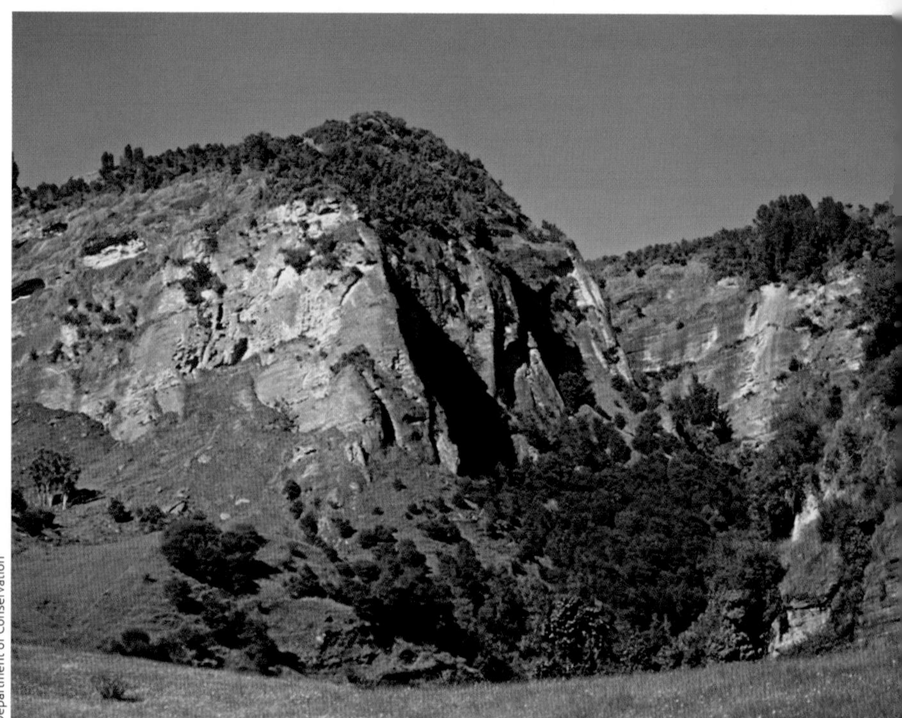

Department of Conservation

*Boundary Stream*

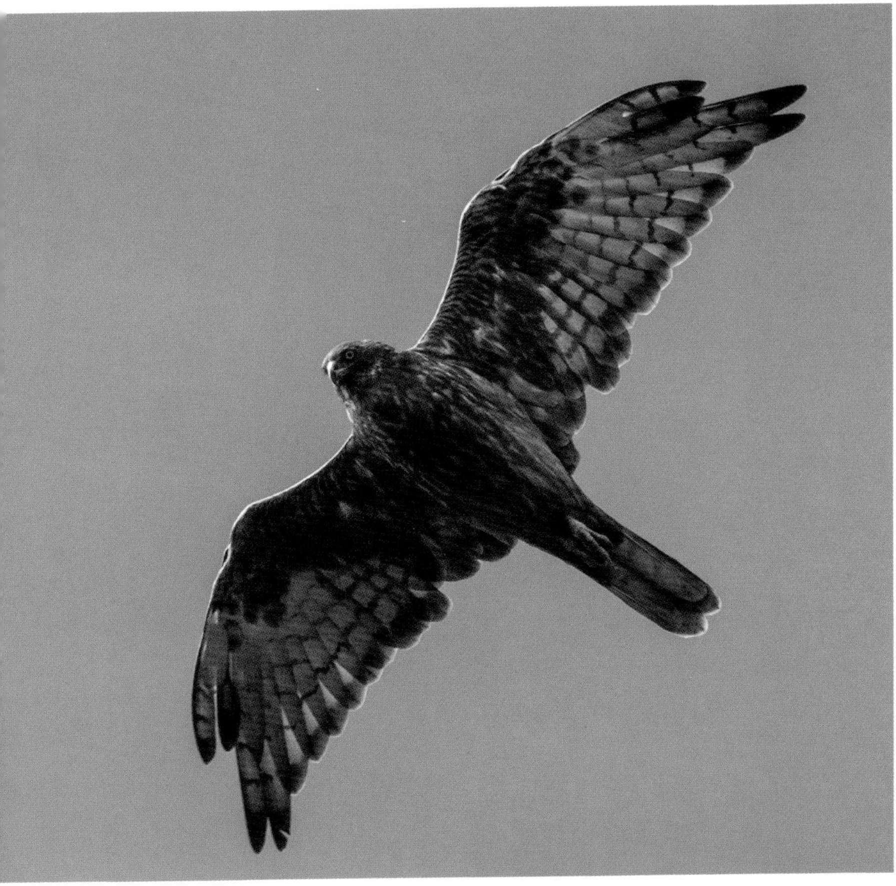

Oscar Thomas

## ustralasian Harrier

*ircus approximans gouldi*

wamp harriers, commonly known as awks, are abundant raptors. They are ften seen on roadkill, and perusing prey n pastureland, gliding elegantly with slow eats of their long wings. The Australasian arrier is big (50–60cm and weighing up o 850g), with long legs and taloned feet, hich tuck under the tail when flying. It as a golden eye, hooked beak, greyish-rown feathers and long wings. It self-ntroduced from Australia in the 1860s and as benefited hugely from the destruction f forest and conversion of it to pasture. It sually nests on the edges of swamps in eeds or rushes. Seen here on the edges of he forest and farmland.

Oscar Thomas

Nir Ketraru

### Kākāriki/Yellow-crowned Parakeet

*Cyanoramphus auriceps*

Yellow-crowned Parakeets are small, bright green, noisy, canopy-dwelling parakeets that can be seen and heard chattering while flying in groups. There are five other similarly sized green parakeets in their genus. They were previously found in most forests in New Zealand but have been decimated by rats, stoats, weasels and possums, particularly while nesting in holes in trees. They are still found in most of the predator-free offshore islands, and numbers are probably in the tens of thousands. They have been successfully translocated to Boundary Stream and do well in this area because of intensive predator control.

### Tauhou/Silvereye

*Zosterops lateralis*

The Silvereye, also known as a wax-eye and tauhou, is a small bird (weighing 12–14g) and a relatively new arrival. It self-introduced, presumably from Australia, in the 1850s, probably blown to New Zealand on the strong predominant west winds. It is now common. Silvereyes have mainly olive-green plumage above and cream below, and a pale yellow head. They are easy to identify because of the white ring around their eyes. They utter a busy little twitter when foraging together, and the male has a melodious, high-noted song during the breeding season. They are omnivores but enjoy fruit, nectar and seeds.

## Accommodation

There are hundreds of accommodation options in Napier, including backpacker facilities, bed and breakfasts, motels, hotels and exclusive lodges. There are possible homestays closer to the area. Google the usual websites.

# Blowhard Bush Scenic Reserve

'he 63ha reserve is rich in natural and social history. The ›rest contains many caves, fascinating rock formations and vater-cut passages. Despite being repeatedly felled for timber, everal forest giant trees still exist, including one 25m mataī hat is 700–800 years old. At the other end of the scale there ; a small, delicate blue orchid living epiphytically on many of he trees. The various walks in the reserve take you through :ānuka and beech forest and many interesting rock passages. 'he area is well known for its limestone caves and bluffs. 'orest & Bird volunteers work on the reserve, planting native pecies and eradicating introduced pests including possums, ats, mustelids, feral cats, and plants such as broom and wilding ›ines. Their efforts have been rewarded with flourishing ›irdlife and regenerating native bush. This information was :xtrapolated from the Forest & Bird website, see: www. orestandbird.org.nz/projects/blowhard-bush-reserve-ıastings. There is a useful PDF brochure with maps at the ›ottom of the link.

## KEY FACTS

**Getting There**
Blowhard Bush Reserve is located at the corner of Lawrence Road and Napier-Taihape Road approximately 53km north-west of Hastings.

**Facilities**
The reserve contains a picnic shelter, two information boards and two toilets.

**Best Time to Visit**
There is good birding in all seasons but more bird activity and song in spring. Tracks can be muddy in winter.

*A forest walk at Blowhard Bush Scenic Reserve*

## Tracks and sites

The reserve contains four well-maintained walking tracks. The **Tui Track** (2½ hours), the **Troglodyte Track** (30–45 minutes), the **Rewi Track** (30 minutes) and the **Rakaunui Track** (45 minutes). There is an excellent lookout towards the Kaweka Ranges.

### Key Species

Korimako/Bellbird, tūī, pīwakawaka/New Zealand Fantail, Grey Warbler, Whitehead, kererū/New Zealand Pigeon, Shining Cuckoo, Long-tailed Cuckoo, Tomtit, Fernbird, New Zealand Pipit, ruru/Morepork, tauhou/Silvereye, North Island Robin and more.

Jan Bottema

*Kererū*

## lew Zealand Tomtit

*etroica macrocephala*

his species has five subspecies on different lands. It is a wee charmer that weighs just g, has a large head relative to its body and prettily black and cream. It has a tendency perch at an angle on tree bark and has a varbling whistle. These hard-working little irds often produce three broods a year, and his seems to keep their population stable espite predation. Like most birds they do well predator-free areas of forest and scrubland. hey are not shy and you can get within a netre of two of them on forest tracks.

Oscar Thomas

## .ong-tailed Cuckoo

*Jrodynamis taitensis*

'his migratory bird only breeds in New ealand. In the winter months, March–eptember, it disperses to a variety of Pacific slands, from Micronesia in the west to the Pitcairn group in the east. It is often difficult o see it as it prefers to live deep in the forest anopy. It has a harsh screech and its tail is s long as its body in flight. It is the larger of the two cuckoo species in New Zealand, brown on top and paler below with bands nd streaks of pattern all over it. Both sexes ook alike. The species is a brood parasite nd lays eggs singly in the nests of much tinier Whiteheads, Brown Creepers or Yellowheads. The birds are so big that they have difficulty laying in the tiny nests of their host's chicks. The host parents work frantically to feed the ever-growing cuckoo chick, doing so for some weeks after it has outgrown the nest. The decline in the range of the host birds has affected the range of the Long-tailed Cuckoo; it is common in much of New Zealand's remaining forests. It mainly eats invertebrates (including large cicadas) and lizards.

Oscar Thomas

## Accommodation

This is in a mountainous, forested area but there is plenty of accommodation in the cities of Napier and Hastings, both approximately 53km away.

# Cape Kidnappers Gannet Colonies

Liz Light

*Nesting gannets*

It is an extraordinary experience to see 17,000 gannets mating, then nesting, then chick-feeding, and to finally see the chicks stretching their newly feathered wings before flying away. Cape Kidnappers, a rocky promontory on a windy, east-pointing cape, has three gannet colonies quite close to each other. The 13ha Cape Kidnappers Reserve includes the Saddle and Black Reef gannet colonies. Both are closed to the public, but the Black Reef colony can be viewed closely from the beach. The Plateau colony (5,000 birds) above the beach on an elevated headland is the best place for viewing the nesting gannets and there are also amazing panoramic views of the sea and rocky islets around the cape. This colony is located on private land. Visitors are asked to cooperate with the landowners by keeping to the defined track and not disturbing the birds or the farm animals.

### Accommodation

There are plenty of places to stay at Clifton and nearby Te Awanga, ranging from camping (Te Awanga camping grounds) to glamping, regular hotels in nearby Napier, and luxury lodges in wineries and farms. Try booking.com, Airbnb and accommodation sites such as bookabatch.co.nz.

## KEY FACTS

### Getting There

Clifton, the village nearest to the cape, is 23km south of Napier, the largest city in the east coast province of Hawkes Bay. It is then 8.5km to the cape.

1. It is a 8.5km walk along the beach from Clifton to Cape Kidnappers and back the same way. It takes five hours and is beautiful. However, you need to check the tides as the beach is impassable at high tide. Google NZ metservices tides Cape Kidnappers and you will find the daily tide times. This walk is free. Take plenty of drinking water and food. DOC has an informative website. Google DOC Cape Kidnappers walking track and you will find it. www.doc.govt.nz/parks-and-recreation/places-to-go/hawkes-bay/places/cape-kidnappers-gannet-reserve/cape-kidnappers-walking-track.
2. Gannet Beach Adventures: www.gannets.com. Sitting on a large trailer and being towed along the beach by a vintage tractor is a fun and a uniquely New Zealand way to visit the colony. The tractor departs from Clifton Beach but the company will pick you up from the towns in the area beforehand. There is an informative commentary along the way and some fascinating geographical features. The trip is four hours return. Departure times are tide dependent. The price was $50 NZ at the time of writing. Along the beach watch for White-fronted Terns, Variable Oystercatchers, Reef Herons, Caspian Terns, various gulls and kororā/Little Penguins.
3. Gannet Safaris Overland: gannetsafaris.co.nz. This is for people who do not like walking or riding on trailers. A comfortable four wheel-drive bus goes to the colony over private farmland. It has the convenience of not being tide dependent but lacks the ambience of options one and two above. The bus departs the Te Awanga base at 9.30 a.m. and returns at 1.30 p.m., and visits two of the three colonies. The price at the time of writing was $88 per person (adult).

### Facilities

Toilets only.

### Best Time to Visit

The gannets are here from August until April.

In August/September they are returning, finding mates with elaborate courtship rituals, squabbling over nesting spaces, and gathering grass and seaweed to help construct the nests.

In November/December they are sitting on eggs.

In December/January the chicks are being fed.

In March/April the chicks are fledging. Then by the end of April most of the birds have gone.

# Pūkaha Mount Bruce National Wildlife Centre

*At the top of Mt Bruce looking west towards the Tararua Ranges*

## KEY FACTS

### Getting There

Pūkaha is on SH2 and is 30km (20 minutes' drive) north of Masterton and 10km south of Eketahuna. It takes about two hours to drive from Wellington or Napier and about one hour from Palmerston North. Open 9 a.m. to 6 p.m. every day except Christmas Day.

### Fees

Adults $20.00, children $6.00, family pass $50, under fives free. There are guided tours at set times for various prices, including a night tour, which gives a good chance of seeing a kiwi in the wild. See the website.

### Facilities

There is an excellent cafe, an interpretation centre, a small but tasteful shop and toilet facilities.

### Best Time to Visit

The kiwi house and aviary make this place is a must-visit destination in all seasons though there is more bird activity and song in spring.

### Useful Website

www.pukaha.org.nz info@pukaha.cog.

Pūkaha Mount Bruce, a bird sanctuary on 942ha of mountainous podocarp forest, is a birder's must-visit destination. Highlights are the kiwi house, the large open-air aviary, the takahē that graze on the cafe lawn and the kākā that swoop in from the forest at feeding time. Pūkaha Mount Bruce is a partnership between the National Wildlife Centre Trust, Rangitaane o Wairarapa and DOC. A key role is the captive breeding of threatened wildlife, undertaken at the National Wildlife Centre, within the sanctuary. Some of New Zealand's most endangered birds are bred here to be released into the wild. Kiwi, kōkako, kākā and other wildlife are being restored to this and other forests. The kiwi house is a place where you can see these nocturnal birds during the day. There is also a large walk-in aviary where you can sit and watch many bird species closely.

## Tracks and sites

**The kiwi house** is dark inside with infrared lighting, so you can see the birds doing their daily activities. Manukura, a white kiwi, is a star. She is not an albino but a genetic throwback to kiwi that may have previously been white. There are tuatara and various endemic geckos and skinks happily living in the kiwi house too.

The large, walk-in and around, **open-air aviary** is an excellent place to see some rare birds in a semi-natural habitat.

A walk follows the **Mt Bruce Stream** for about 300m. Kererū/New Zealand Pigeons drink from the water, tūī zoom between branches chasing each other and pīwakawaka/New Zealand Fantails flit about. Tiny Grey Warblers and North Island Robins fidget in the trackside undergrowth, and in the nesting season Riflemen whizz in and out of next boxes.

### Key Species

Paradise Shelduck, Northern Brown Kiwi, korimako/Bellbird, Grey Warbler, pīwakawaka/ New Zealand Fantail, takahē, Whitehead, tīeke/ North Island Saddleback, kererū/New Zealand Pigeon, kākā, tūī, Tomtit, ruru/Morepork, Grey Warbler, Rifleman, kōkako, pūkeko, tauhou/ Silvereye, Shining Cuckoo, Long-tailed Cuckoo, Sacred Kingfisher, Welcome Swallow, Little Shag, Australasian Harrier, kārearea/New Zealand Falcon.

**Kākā are fed once a day in a glade** (mid-afternoon). These large, attractive parrots fly in from the surrounding bush and can be seen en masse at this time. Spectacular.

**There is an excellent bush walk** to the lookout on the top of Mt Bruce on a well-maintained track. You can walk up to the lookout and back, or complete a loop that is 4km long and takes about two hours. Most of the bird action is lower down, near the stream and the visitor centre.

Liz Light

*Kākā*

Oscar Thomas

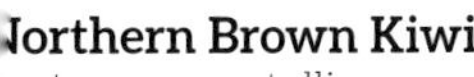

### Jorthern Brown Kiwi

*\pteryx mantelli*

his bird is flightless and nocturnal, and as whiskers, nostrils at the end of its long eak and long, strong legs. It has dark brown eathers streaked with reddish and black hades. The female lays one huge egg annually, he largest egg in the world in proportion to ody size; some 20 per cent of it. Kiwi nest in urrows, holes in trees and other sheltered avities. The chicks hatch fully formed and nly stay in the nest for five days, then are ut foraging with their parents. They can live or 50 years and grow to 25–40cm tall. Being octurnal, kiwi are heard more than they are een, and the male has a trilling, warbling cry vith rising notes. The female is quieter and nore gruff, and the birds sometimes duet. Despite being the most common kiwi species, lue to habitat loss, and predation by dogs, ats and stoats, the population of this species, estimated to be up to 21,000, is declining in reas without predator control. It is thriving in he sanctuary.

### Rifleman

*Acanthisitta chloris*

The Rifleman weighs about 7g and is New Zealand's smallest bird. It is found in parts of the North Island and most of the forested areas of the South Island. The upper body is bright green on the male and yellowy-green on the female, and it has a tiny tail. It is monogamous and builds oval nests in cavities like hollows in trees. In Pūkaha there are nest boxes for the Rifleman alongside some of the tracks – it is pure pleasure to see these fast, tiny things whipping in and out and, in season, hearing the cheeps of their miniature chicks.

## Accommodation

Motorhomes can park for the night here (for an additional fee), and there is an electric car charging facility. There is ample accommodation of different kinds in Masterton. There are also excellent cafes and a boutique brewery 15 minutes' drive away.

# Waikanae Estuary

Laurie Petherick

*Waikanae Estuary*

This reserve, at the mouth of the Waikanae River, protects a natural medley of small freshwater lakes, brackish lagoons and marshes, tidal sandflats and sandy beaches. It comprises 75ha and is bounded by two sandpits, to the north (Waikanae) and the south (Paraparaumu). Here, freshwater from the Tararua Ranges meets the sea. This mixing of the waters and the ever-shifting river mouth creates an environment sustaining rich plant and animal communities. It is an important habitat for birds and more than 60 bird species breed here, including the Banded Dotterel, pūkeko, New Zealand Dabchick and Variable Oystercatcher. They occupy the sandspit where the river affords some protection from the increasing numbers of cats, ferrets, dogs and trail bikes on the other side of the reserve. White-fronted Terns, shags, gulls and penguins roost on land but rely on the sea for small fish and shrimps. Domestic migrants, such as Black-fronted Terns and Wrybills from the South Island, often visit, as do northern hemisphere migratory birds like godwits and knots. The birds depend upon the constant movement of the river and the tide, and the tidal flats are revealed twice a day with their bounty of worms, shellfish and crabs.

Waikanae Estuary Care Group is long established and is continually improving the estuary with planting, predator and weed control, and track maintenance. It keeps a watch for vehicles like bikes, and dogs that stray illegally on to the beach and may damage nests and disturb birds. The reserve is challenged by being hemmed in by the suburbs of

## KEY FACTS

**Getting There**

The Waikanae Estuary is located between Paraparaumu and Waikanae on the Kāpiti coast 60km north of Wellington. See brochure and map for details of access: www.gw.govt.nz/assets/WRS/Biodiversity/37-Waikanae-Estuary.pdf. Get a local guide: Mik and Moira Peryer are locals and passionate about birds. They can take you on a personal birdwatching tour of the estuary. Tel.: +64 4 905 1001, Mobile: 021 750 603, email: michaelperyer@gmail.com.

**Facilities**

The nearby town of Waikanae has public toilets, parks, cafes and shops.

**Best Time to Visit**

October to March to see the migratory wading birds.

aikanae town, which adds issues relating drainage and the risk to birds of predation domestic cats and dogs.

## racks and walks

ıere are 5km of tracks, all flat and easy alks. A main entry point to the tracks is off anly St North, Paraparaumu Beach. One ack leads out on to the seaward side of the goon and the other follows the inland side the estuary. There is also a car park at the ıd of Tutere Street, on the Waikanae side the estuary, which joins the track system, ıd Otaihanga Domain off Makora Road also ıs good access to the tracks.

### Key Species

Black Swan, Paradise Shelduck, Grey Teal, Australasian Shoveler, New Zealand Scaup, Little Shag, Black Shag, Pied Shag, White-faced Heron, Royal Spoonbill, Swamp Harrier, pūkeko, Bar-tailed Godwit, Lesser Knot, Variable Oystercatcher, tōrea/South Island Pied Oystercatcher, South Island Pied Stilt, Banded Dotterel, Spur-winged Plover, Arctic Skua, Southern Black-backed Gull, Red-billed Gull, Black-billed Gull, Caspian Tern, Black-fronted Tern, White-fronted Tern, Sacred Kingfisher, Fernbird, Grey Warbler, Welcome Swallow.

Oscar Thomas

*Sacred Kingfisher*

Gert op den Dries

*White-fronted Tern*

Oscar Thomas

## Spur-winged Plover
*Vanellus miles*

The Spur-winged Plover (or Masked Lapwin, self-introduced from Australia 80 years ago and is now all over New Zealand in open pastureland, riverbeds and urban parks. It ha a harsh, shrill call and annoys horticulturalis by damaging leafy green vegetables. It is not loved but should be admired. It is successful because it is aggressive, vigorously, noisily and successfully defending its chicks against predators. It is well adapted to grassland and enjoys the grassland and planted areas of the Waikanae Estuary.

## White-fronted Tern
*Sterna striata*

White-fronted Terns are endemic and breed around most of the coast on rock stacks, beaches, dunes, riverbeds, and the piles of bridges and wharfs that are no longer used. They are medium sized, and have a long tai and a black cap and bill. They often roost together at high tide on wharfs and moored boats. Most remain in New Zealand for the winter.

Oscar Thomas

### Accommodation

Waikanae is a pleasant coastal town and a holiday destination for people from Wellington and Palmerston North cities. There is plenty of accommodation available, from camping and cabins, to motels.

# Kāpiti Island Nature Reserve

The bush, beach and big skies of Kāpiti Island

A visit to Kāpiti Island is one of the best possible birding experiences. It is a paradisical, tranquil, predator-free nature reserve on a relatively large, densely forested island. It is possible to get close to rare birds such as kākā, kōkako, takahē and hīhī/Stitchbird, and people who stay overnight have a very good chance of seeing two species of kiwi.

Kāpiti Island lies about 8km off the west coast of the lower North Island. It is 10km long and 2km wide, covering an area of 1,965ha. The highest point, Tuteremoana, is 521m above the sea, and there are superb views over land and ocean at the top. The island is a semi-closed sanctuary and can only be accessed by approved tour operators. Casual, private boats cannot land there.

Superb birding off the west coast of the lower North Island

## KEY FACTS

### Getting There

You must go with an approved tour operator. There are two, both good with comparable prices.

1. Kāpiti Island Eco Experience, Tel.: 0800 433 779, www.kapitiislandeco.co.nz. Transport only to Rangatira Point, adult $80, child (5–17 years) $40, under five free. Wellington return. Includes pick-up/drop-off from Wellington hotels and a picnic lunch, adult $185, child (5–17 years) $145.
2. Kāpiti Island Nature Tours, Tel.: 0800 527 484 or 021 126 7525, www.kapitiisland.com. Transport only to Rangatira Point, adult $82, half that for a child, under five free. Overnight kiwi spotting, including two full days, glamping or staying in a cabin, all meals and fully guided, adult $393, child $230. Both tours depart each day (weather permitting) from the Kāpiti Boating Club, Kāpiti Road Paraparaumu Beach, Paraparaumu. There is an excellent birding checklist and map: www.kapitiisland.com/library/bird-checklist.pdf.

### Best Time to Visit

All seasons have good birding but check the weather closely in winter. If the sea is rough, the boat will be cancelled.

## Tracks and sites

Keep a lookout for seabirds (and dolphins) on the boat journey to the island. There are two tracks on the island. The **Trig** and **McKenzie Tracks** lead up to the summit and join about three-quarters of the way up to become a single track to the top. It takes about three hours to walk to the summit and back. The tracks are steep in places but are well maintained. There are smaller bush and beach walks in the area of Rangitira Point, where the boats arrive and depart from in the middle of the island.

Liz Light

*Pied Shag*

### Key Species

Little Spotted Kiwi, Northern Brown Kiwi, takahē, pūkeko, Little Blue Penguin, Fluttering Shearwater, tītī/Sooty Shearwater, Australasian Gannet, Black Shag, Pied Shag, Little Shag, Spotted Shag, Black Swan, Grey Duck, pāteke/Brown Teal, Paradise Shelduck, White-faced Heron, Reef Heron, Royal Spoonbill, North Island Weka, Variable Oystercatcher, Shore Plover, Arctic Skua, Black-backed Gull, Red-billed Gull, Black-billed Gull, Caspian Tern, White-fronted Tern, Black-fronted Tern, poaka/Pied Stilt, Bittern, kererū/New Zealand Pigeon, kākā, kākāriki/Red-crowned Parakeet, Long-tailed Cuckoo, Shining Cuckoo, ruru/Morepork, Sacred Kingfisher, Grey warbler, tīeke/North Island Saddleback, Tomtit, North Island Robin, hīhī/Stitchbird, korimako/Bellbird, tūī, pīwakawaka/New Zealand Fantail, New Zealand Pipit, Whitehead, tauhou/Silvereye, Welcome Swallow, Fernbird, kārearea/New Zealand Falcon, Australasian Harrier.

Oscar Thomas

*Fluttering Shearwater*

## ittle Spotted Kiwi

*pteryx owenii*

ne Little Spotted Kiwi is the smallest of the ve kiwi species, standing at about 30cm. It as extinct on the mainland by 1980, but population on predator-free Kāpiti Island thriving (more than 1,200 and growing). anslocations to other offshore islands and edator-free inland sanctuaries have allowed ne numbers of this little kiwi to increase. he species is nocturnal and flightless, but an easily be seen by birdwatchers who stay ver on Kāpiti Island. It is light browny-grey nottled or banded horizontally, with a long ale bill, and short pale legs, toes and claws.

Sallie Bassett

## ied Shag

*halacrocorax varius*

he Pied Shag is a large, black-and-white ird 65–85cm in length and weighing p to 2kg. It can be found around much f the coastline. It is impressive looking, vith blue eyes, a yellow patch on its head and primarily back topsides with white undersides from the beak along the throat and belly. It is seen individually or in small groups roosting on rocky headlands or trees. The birds are colonial nesters with untidy nests in trees near the coast. They feed on fish up to 15cm long, and some crustaceans.

Glenda Peake

### Accommodation

Paraparaumu is a pleasant coastal town and a holiday destination for folk from Wellington and Palmerston North cities. There is plenty of accommodation available from camping and cabins to motels.

# Zealandia Eco-sanctuary

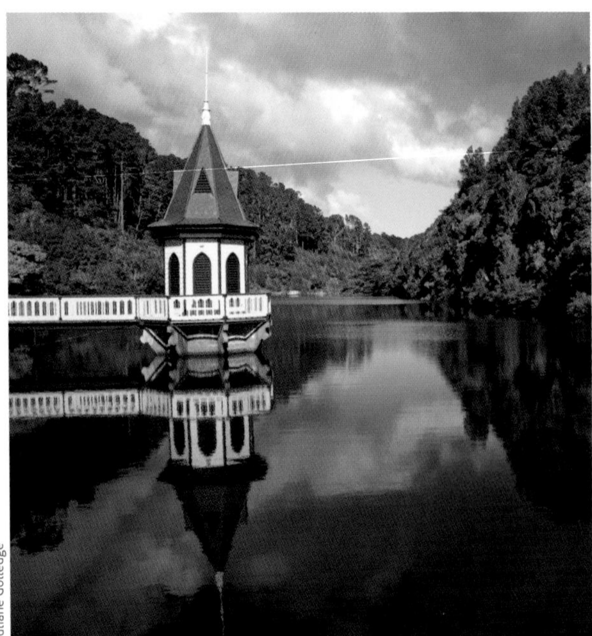

Juliane Golledge

*The lake and valve tower of Zealandia, a predator-free sanctuary and one of Wellington's water reservoirs*

Zealandia is a protected forest and bird sanctuary very close to the centre of Wellington city. The 225ha sanctuary retained its forest as a water catchment area for the city and in 1999, after a huge conservation effort, the predator-proof fence around it was completed. Since then Zealandia has become a safe haven for a number of endangered bird species that fly (and roam for the flightless ones) freely through the valley. The steep, bush-covered terrain is complemented by two lakes and a stream, which adds to the diversity of the bird habitats. There are terrestrial forest birds, waterbirds and much more. Zealandia is easy to access and gives an impressive introduction to many New Zealand birds in the wild. This place makes birdwatching easy and is a fine example of what concerted and determined conservation efforts can achieve. There is a range of tracks, and bird enthusiasts can easily spend a full day here; there are more than 40 indigenous/endemic species to see and hear. There is an excellent interactive exhibition centre with lots of information, and the Rātā Cafe is a great place to refuel between walks.

## KEY FACTS

Getting There

Driving: Zealandia is close to Wellington CBD, so getting here is easy. 53 Waiapu Road, Karori, Wellington. There is limited free parking available at the visitor centre. Additional free parking is available 500m away in the Birdwood Road car park.
Free shuttle: there is a free shuttle bus from outside the Wellington city i-SITE and top of the cable car. See website for timetable.
Public bus: many of Wellington's public buses will stop at the end of Waiapu Road. This is about a two-minute walk from the sanctuary. See www.metlink.org.nz for timetable.
Cable car: take the Wellington cable car from Lambton Quay to the top of the Botanic Gardens. From there it is a 20–30 minute walk. See www.wellingtoncablecar.co.nz.
Zealandia is open from 9 a.m. to 5 p.m. every day except Christmas Day.
**Fees** Adults $19.50, children to 17 $10.00, family pass $46, under fives free. There are guided tours at set times for a variety of prices. See the website.

Facilities

Rātā Cafe is open at weekends. There is an an interpretation centre, a small but tasteful shop and toilet facilities.

Best Time to Visit

This place is a must-visit destination in all seasons though there is more bird activity and song in spring.

Useful Website

www.visitzealandia.com is an excellent website full of information and a track map.

## racks and tours

here is an excellent map at the centre nd tracks and bird-feeding stations are arked. There are numerous tracks, from ng, steep ones that cover a variety of rrain, to shorter strolls. The tracks are well aintained but can nonetheless be muddy fter rain so wear decent walking shoes. here are guided tours during daytime, wilight and night. See website for costs and mes. Visitors can walk freely on the tracks nd stay all day.

### Key Species

Ruru/Morepork, hīhī/Stitchbird, North Island Robin, tīeke/North Island Saddleback, takahē, tūī, pīwakawaka/New Zealand Fantail, tauhou/Silvereye, korimako/Bellbird, New Zealand Pipit, Little Spotted Kiwi, Shining Cuckoo, Australasian Harrier, kākāriki/Red-crowned Parakeet, kārearea/New Zealand Falcon, Grey Warbler, Sacred Kingfisher, kererū/New Zealand Pigeon, Welcome Swallow, Whitehead, kākā, pāteke/Brown Teal, Grey Duck, New Zealand Scaup, Black Swan, Welcome Swallow, Southern Black-backed Gull, Black Shag, Little Shag, Little Black Shag, Pied Shag, Dabchick, Paradise Shelduck, Royal Spoonbill.

Oscar Thomas

*Ruru*

Justine Hall

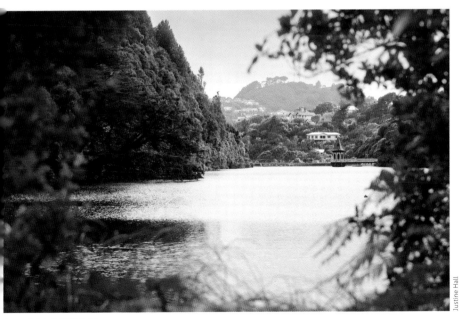

*Lower Lake at Zealandia*

## Whitehead

*Mohoua albicilla*

Whiteheads are tiny, delicate, insect-eating birds. They are busy and gregarious birds, often tweeting, chirping and trilling together in large groups. Males, which weigh up to 20g and are half the size of sparrows, have white heads and bellies, while their wings and tails are light brown. Females are slightly smaller and have a brown crown and nape. Whiteheads are only found in the North Island and are flourishing in places like Zealandia with predator control. They can be seen throughout the sanctuary, in small groups high in trees. They are usually seen looking for spiders and insects on tree trunks and leaves, often hanging upside down as they search for food.

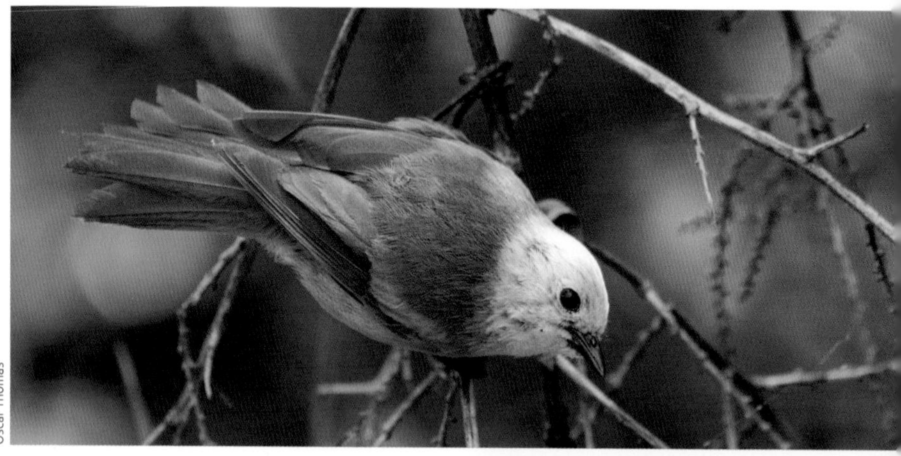

Oscar Thomas

## Little Shag

*Phalacrocorax melanoleucos brevirostris*

Little Shags began nesting at Zealandia in 2003. Numbers roosting at the sanctuary fluctuate seasonally, with up to 70 birds during winter and fewer than 10 in February, when nesting is over and fledglings are independent. They are often seen roosting and nesting on the large macrocarpa tree on the western side of the lower lake and other smaller trees nearby. Shags are efficient underwater swimmers and feed on fish and small crustaceans. After fishing they spend time perched with their wings spread, drying their feathers. Their feathers are not waterproof, making it easier for them to dive and stay underwater, but this means that the birds quickly get waterlogged and cold and must dry off afterwards.

Oscar Thomas

### Accommodation

It is in central Wellington, the capital city, so there are endless accommodation options.

# South Island

1 Wairau Lagoons
2 Kaikōura for Pelagic Birds
3 Waimea Inlet
4 Kahurangi National Park
5 Farewell Spit
6 Paparoa for Westland Petrels
7 Ōkārito Lagoon and Ōkārito Kiwi Sanctuary
8 Waitangiroto Nature Reserve, kōtuku/White Heron Colony
9 Monro Beach
10 Makarora, Mt Aspiring National Park
11 Avon–Heathcote/Ihutai Estuary
12 Lake Ellesmere/Waihora
13 Arthur's Pass National Park
14 Tasman River and Black Stilt
15 Oamaru Blue Penguin Colony
16 Royal Albatross Centre, Pukekura, Taiaroa Head
17 Otago Peninsula and Yellow-eyed Penguin
18 Fiordland National Park
19 Waituna Lagoon
20 Rakiura/Stewart Island
21 Ulva Island
22 Chatham Islands
23 Pitt Island

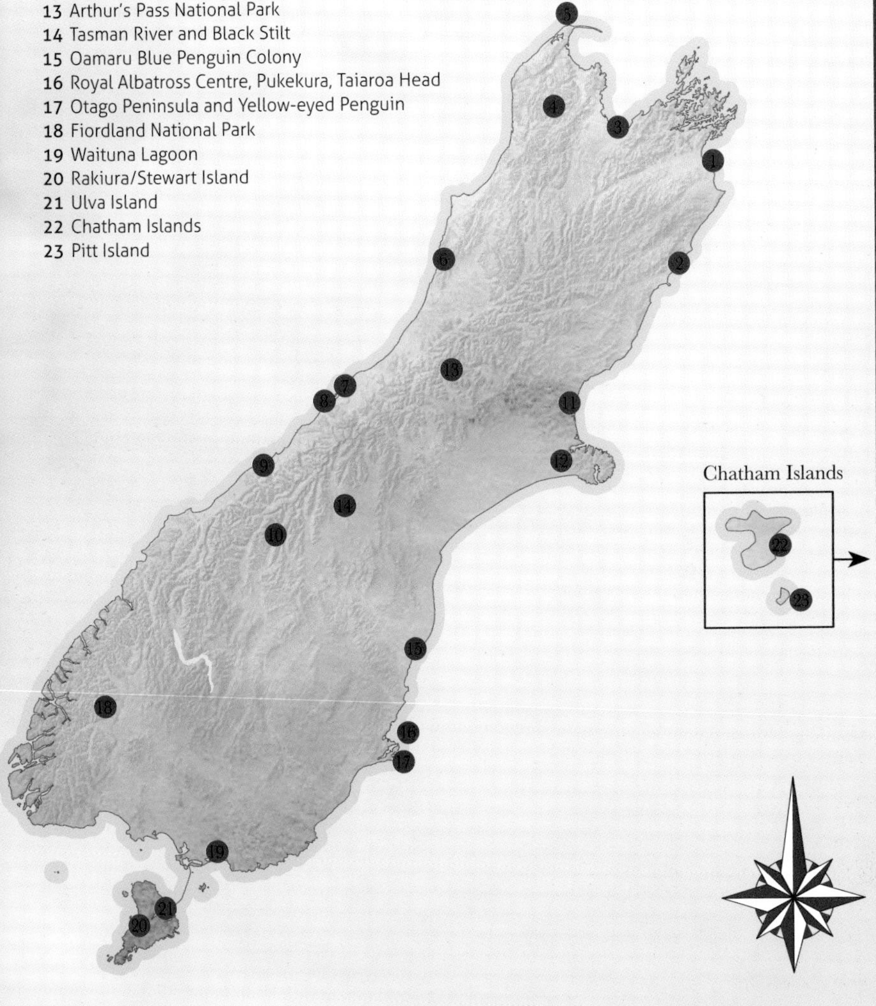

Chatham Islands

# South Island overview

It is 900 km from the top (latitude 40.5 south) to the bottom of the South Island (46.45 south). The Southern Alps stretch for 700 km north-east to south-west. The highest is Aoraki/Mount Cook in the middle of the Alps, and there are 16 other mountain peaks over 3,000 metres. It's a substantial range of mountains and it cuts across the prevailing weather pattern, which means there is heavy and regular rainfall in the west, and rain-shadow areas to the east of the Alps. The West Coast rainforest is spectacular as are many of the birds that live in it.

Unique geographical features pertaining to birds include many braided rivers, unravelling as they travel from the mountains across the plains to the sea. They are unusual ecosystems of international importance and provide habitat for several endemic threatened species, for instance, Wrybill, Banded Dotterel, Black-fronted Tern, Black-billed Gull, and kakī/Black Stilt.

Farewell Spit, 11,380 hectares of bird sanctuary, at the top of the South Island, is recognised under the RAMSAR Convention as being vital for shorebirds. It's especially important for migratory shorebirds on the East Asian–Australasian Flyway. These birds nest in the Arctic but return to the Spit for the summer months. 10,000 Red Knots and 12,000 Bar-tailed Godwits are some of the many bird species that can be seen here from September to March.

## Iconic bird

**Bar-tailed Godwit** Each March godwits gather near tidal mudflats prior to their long journey to Alaska and godwit-fanciers also gather to farewell them and wish them a successful return in spring. Godwits are exceptional long-distance fliers and have been recorded as doing 11,000 km, in nine days, without stopping. They make their way to the tundra areas of western Alaska to nest and return again every spring. Godwits are often a metaphor for young New Zealanders, who leave home for work and adventure in different parts of the world, but usually, eventually, return.

Oscar Thomas

The South Island is also home to some extraordinarily unusual avian species, such as the kea, the world's only alpine parrot, and the kākāpō, a large, flightless, nocturnal, ground-dwelling parrot. The takahē, the largest flightless rail in the world, was considered extinct for 50 years until a few were discovered in a remote valley 1948.

Liz Light

*Arthur's Pass area features braided rivers, beech forest, tussock grassland and alpine vegetation.*

# Wairau Lagoon

ew from Wairau Lagoon walkway looking south

/airau Lagoon is east of the town of Blenheim in the 1arlborough region in the north of the South Island. comprises 2,000ha of lagoons, rivers and wetlands here the Wairau and Opawa Rivers are prevented from ntering the Pacific Ocean by the Wairau Bar. The lagoon separated from Cloudy Bay by an 11km spit, made of nall boulders, known as Pokohiwi. To the south, the Vhite Bluffs soar to 268m, and at the northern end of the oit lies a sacred Moa Hunter burial site. Within this vast vetland area there are several key birding sites, and some 0 species, including wetland and wading birds, waterbirds nd seabirds, are seen in the lagoons. For birders there are our main areas of interest: the Wairau Lagoon Walkway, ncluding the sewage-treatment ponds, the Grovetown agoon, Taylor Dam Reserve and the Wairau Bar. Note that his is a major wine-growing area and arguably the world's est Sauvignon Blanc comes from the nearby vineyards. njoy a delicious tipple after birding.

## KEY FACTS

**Getting There**
Turn off State Highway 1 at Grovetown. Proceed through suburban Grovetown (Fell Street), cross Vickerman Street, and keep going. Before the end of the road turn right into Steam Wharf Road. Halfway down Steam Wharf Road you can see the lagoon and a place from which you can launch kayaks. You can park at the end of Steam Wharf Road.

**Facilities**
Picnic tables and bird hides.

**Habitat**
Intertidal river flats, tidal channels, beach, wetlands and marsh, forest-covered islands, sewage ponds.

**Best Time to Visit**
October to March for the migratory wading birds and spring for the spoonbills nesting.

## Tracks and sites

**Wairau Lagoon Walkway**, marlboroughnz.com/guides/walks/wairau-lagoons-walkway. There is a loop track which takes about three hours to walk. The path is sometimes wet and muddy so wear suitable boots. On the walk are the sewage ponds, and 80 pairs of spoonbills nest on an island in trees. This is the biggest nesting colony of spoonbills in New Zealand. In spring you can see them displaying their glamorous mating plumage. Glossy Ibis, recently self-introduced from Australia, also nest on this island among the spoonbills. Locals recently sighted 18 birds and six nests. To get there, drive to the end of Hardings Road, which leaves State Highway 1, 1.5km south of Blenheim. See a map on the above website.

**Grovetown Lagoon walk**, www.grovetown.co.nz/About-The-Lagoon/visit-the-lagoon. Grovetown Lagoon is an oxbow loop in the Wairau River. There is a 3.8km circuit track around the outside of the oxbow loop, leaving birdlife undisturbed on the inside of the bow. Birds commonly seen include pūkeko, Australasian Shoveler, Paradise Shelduck, Grey Teal, New Zealan Scaup, Australasian Bittern, Black Swan, Black Shag, Little Shag, Black-backed Gull, Welcome Swallow, Sacred Kingfisher, and Harrier Hawk.

**Wairau Bar** On the north side of the lagoo Wairau Bar Road almost reaches the river mouth. This is a good area for seabirds and shags, including the Spotted Shag. Watch o for Black-fronted and Caspian Terns worki up and down the river mouth, and for Bar-tailed Godwits and Golden Pacific Plovers i November–March on the estuary side of th bar. There is also a wetland to the north of the Wairau Bar, and Australasian Bitterns and poaka/Pied Stilts can be seen here.

**Taylor Dam** Off Taylor Pass Road, Taylor Dam, only five minutes from Blenheim, was built in 1965 to control flooding and has become a fine bird sanctuary. Besides the usual wetland and waterbirds the New Zealand Dabchick can be sighted. There are picnic tables and this is a quiet place to stay for self-contained camper vans. There are n toilets so no freedom campers.

Oscar Thomas

*Royal Spoonbill*

*esting Glossy Ibis on the artificial islands at the Blenheim oxidation ponds*

*Ruddy Turnstone*

## Key Species

Black Swan, Paradise Shelduck, Australasian Shoveler, Little Shag, Black Shag, Pied Shag, Spotted Shag, kōtuku/White Heron, White-faced Heron, Royal Spoonbill, Australasian Bittern, Swamp Harrier, pūkeko, Bar-tailed Godwit, Lesser Knot, Sharp-tailed Sandpiper, Ruddy Turnstone, Pacific Golden Plover, Banded Dotterel, Black-fronted Dotterel, Spur-winged Plover, tōrea/South Island Pied Oystercatcher, South Island Pied Stilt, Variable Oystercatcher, Black-backed Gull, Red-billed Gull, Black-billed Gull, Caspian Tern, Black-fronted Tern, White-fronted Tern, Sacred Kingfisher, korimako/Bellbird, pīwakawaka/New Zealand Fantail, Fernbird, tauhou/Silvereye, Welcome Swallow, New Zealand Pipit.

Oscar Thomas

## Royal Spoonbill
*Platalea regia*

The spoonbill is a large white bird with a distinctive long, black, spoon-shaped bill and black legs. It flies with its neck outstretched and its legs trailing. It roosts in flocks and feeds in small groups both day and night, depending on the tide. The breeding plumage includes a crest of long white plumes at the back of the head, and yellowish feathers on the breast. There is a nesting colony of some 80 pairs on an island, in ngaio trees, in the sewage-treatment ponds on the Wairau Lagoon Walkway.

## Spotted Shag
*Stictocarbo punctatus*

The Spotted Shag is the most handsome of the shag family. It has two black Mohican-type crests on the top of its head, and bright blue-green skin that stretches from its blue-surrounded eyes to its long peachy bill. It has two wide, curved white stripes running from its eyes down either side of its neck and, during the breeding season, the grey plumage has clear black spots. It is abundant in most of New Zealand and breeds on rocky islands and mainland cliffs. It is a treat for bird photographers and can be found around the mouth of the Wairau Lagoon.

## Accommodation

Will and Rose Parsons, at Driftwood Retreat, on the edge of the Opawa River, have 8ha of wetlands in their 14ha property. They have two eco-apartments set in a garden environment. This is a great place to stay for birders. See www.driftwoodretreatnz.com. The nearby town of Blenheim has a range of accommodation options.

# Kaikōura for Pelagic Birds

ack-browed Albatross

## KEY FACTS

### Getting There

Kaikōura is a beautiful little town hemmed in between snowy mountains and deep blue ocean. It is 180km north of Christchurch (2½ hours driving) and 130km (1 hour 40 minutes) south of Blenheim. Albatross Encounter Kaikōura (www.albatrossencounter.co.nz) offers three daily scheduled 2½-hour tours in summer and two daily in the winter. Full-day tours are available on enquiry. The boat leaves from Albatross Encounter, Kaikōura, on the Esplanade. If you get seasick the local pharmacy can sort that out.

### Best Time to Visit

All seasons provide good birding though what can be seen differs with the seasons. For instance Cape Petrels are in the hundreds over the winter and Hutton's Shearwaters are more often sighted in the spring and summer when they are nesting in the nearby mountains. The winter weather can be staunch and boat trips may be cancelled.

his is one of the best places in the world to see seabirds – eir diversity is excellent and it is a bird photographer's eam. There is nowhere else in the world where you e likely to see so many different albatross species. The Vandering Albatross, Royal Albatross, Black-browed lbatross, Buller's Albatross, Black-browed Albatross, ibson's Albatross, White-capped Albatross and Salvin's lbatross are commonly sighted, as well as, occasionally, her rare species. There is also an abundance of petrels, rions and shearwaters of many species. The presence of ich a large variety and number of seabirds in the ocean off aikōura is due to a deep-water trench, the c Canyon, below ie mountains and close to the shore. Upwellings of cold utrient-rich waters over shallower coastal waters produce healthy and varied food chain, hence a large number of sh species. Marine mammals such as sea lions, seals and a ariety of whales and dolphins also inhabit this area. The asiest way to get a close to seabirds is to go on a tour with lbatross Encounter. Boat trips with experienced skippers nd knowledgeable birders take you out to the deep-water eeding grounds a few kilometres offshore.

Hutton's Shearwaters gather near Kaikōura

## Tours and sites

Kaikōura birding is not just about the Albatross Encounter boat trip. **Point Keane**, at the end of the Peninsula Walkway, has amazing all-round views and great birding, especially on a windy day when the sea is too rough for the boats. Birders need good binoculars or a scope. Besides the birds, the path passes close to a seal colony. Do not approach the seals. **Kaikōura sewage-treatment pond** is pretty and a hotspot fo ducks, including many New Zealand nativ endemic species. **Mount Fyffe Forest** and **Kowhai Bush** are popular birding spots fo the native/endemic passerines, and other forest birds.

Oscar Thomas

*White-capped Albatross*

Oscar Thomas

*Southern Royal Albatross and White-capped Albatrosses*

## ape Petrel *Daption capense* and nares Cape Petrel *D. c. australe*

ne Cape Petrel is a beautifully patterned ack-and-white petrel also known widely the Pintado Petrel. It is seen off Kaikōura roughout the year, with flocks of hundreds om late autumn to early spring. The birds ove north for the winter from around ntarctica and New Zealand's subantarctic ands. They co-feed with whales and fur als, with an eye out for any food scraps. wo subspecies are recognized: the Cape etrel, which breeds on the Antarctic eninsula and Antarctic and subantarctic ands, and the Snares Cape Petrel, which reeds on New Zealand's Snares, Bounty, ntipodes, Auckland and Chatham Islands.

Dennis Buurman Photography

## Hutton's Shearwater *Puffinus huttoni*

This shearwater is the only seabird globally to breed in an alpine environment, and it chooses the Seaward Kaikōura Mountains, at an altitude of 1,200–1,800m, to nest. There are two breeding colonies remaining and this is the only place where the species breeds. There is an estimated breeding population of more than 100,000 pairs, with a similar number of non-breeding birds visiting colonies early in the breeding season.
The two alpine colonies are located in the headwaters of the Kowhai River (Uerau Nature Reserve) and on Puhi Peaks Nature Reserve. Hutton's Shearwater is a small, black-and-white shearwater, 36–38cm in length and with a wingspan of about 75cm. The upperparts are brownish-black. The underwing is off-white with broad brownish borders and grey armpits. The bill is long, slender and dark grey, and the feet are pink with black webbing.

## Key Species

Little Blue Penguin, Wandering Albatross, Royal Albatross, Black-browed Albatross, Buller's Albatross, Gibson's Albatross, White-capped Albatross, Salvin's Albatross, Giant Petrel, Buller's Shearwater, tītī/Sooty Shearwater, Flesh-footed Shearwater, Fluttering Shearwater, Hutton's Shearwater, Short-tailed Shearwater, Common Diving Petrel, tāiko/Westland Petrel, Cape Petrel, Grey Petrel, Antarctic Fulmar, Fairy Prion, Grey-faced Petrel, Australasian Gannet, Black Shag, Pied Shag, Little Shag, Spotted Shag, Arctic Skua, Black-backed Gull, Red-billed Gull, Caspian Tern, White-fronted Tern.

## Accommodation

There is a variety of accommodation in Kaikōura, including camping, cabins, bed and breakfasts, backpackers' lodges and upmarket lodges.

# Waimea Inlet

Liz Light

*Waimea Inlet with the western suburbs of Nelson city in the background*

Waimea Inlet is a shallow, bar-built inlet located within Tasman Bay alongside the city of Nelson, the suburbs of Stoke and the town of Richmond. It covers 34km² of intertidal flats and river and tidal channels, and has an internal coastline of 65km. There are 10 islands, comprising about 300ha in total, which contribute to the habitat diversity. The inlet has two tidal openings at opposite ends of Rabbit Island, the primary barrier between Tasman Bay and the inlet. The tidal range is up to 4.2m and an estimated 62 million cubic metres of water is drained with each tide, giving the inlet a decent flush.

Waimea Inlet is a wetland of national importance for shorebirds, including migratory Red Knots, Bar-tailed Godwits, Variable Oystercatchers, tōrea/South Island Pied Oystercatchers and Wrybills. Over time the ecological purity of the Waimea Inlet ecosystem has been severely degraded by its some 50,000 people. DOC has been working with the wider community to restore and rejuvenate the inlet since 2009 and progress has been made. Restoration benefits include the return of threatened native fish to streams and wetlands, and the re-creation of lost habitat for native bird species such as the Banded Rail and Marsh Crake.

## KEY FACTS

**Getting There**
Waimea Estuary is huge and is on the doorstop of Nelson city. The airport is on the edge of it and you get a bird's eye view flying in. The estuary has 10 islands, three of which are accessible by vehicle or bike by causeways. Besides the islands there are numerous other areas good for birdwatching (see sites).

**Best Time to Visit**
September–March for the migratory wading birds.

tes

you would like to see a Banded Rail ntact the people from The Battle for the nded Rail, which is a part of the Tasman vironmental Trust (www.tet.org.nz).

**ll Island** is good for spotting migratory aders such as the Bar-tailed Godwit, 'himbrel, Sharp-tailed Sandpiper and Red not, on the mudflats during a rising tide the shell banks at high tide. There is a useway to Bell Island for vehicles and cles, but it is under water at high tide.

**ne sewage-treatment ponds** on Bell Island e good for spotting waterbirds such as cks, swans and spoonbills.

ot quite in the Waimea Inlet but just a lometre to the east is **Nelson Haven**. This arbour is separated from Tasman Bay by Te okohiwi/Boulder Bank. The 8km boulder ank is an extraordinary geographical feature, an interesting walk and a place from which to see seabirds, including three species of gull and three of tern. Also, Nelson Haven, inside the Boulder Bank, has its share of migratory waders.

### Key Species

Black Swan, Paradise Shelduck, Grey Duck, Black Shag, Pied Shag, Little Shag, Little Black Shag, Spotted Shag, Cattle Egret, White-faced Heron, Royal Spoonbill, Australasian Harrier Hawk, Banded Rail, Marsh Crake, pūkeko, tōrea/South Island Pied Oystercatcher, Variable Oystercatcher, poaka/Pied Stilt, Spur-winged Plover, Banded Dotterel, Bar-tailed Godwit, Red Knot, Lesser Knot, Ruddy Turnstone, Southern Black-backed Gull, Red-billed Gull, Black-billed Gull, Caspian Tern, White-fronted Tern, Black-fronted Tern, kererū/New Zealand Pigeon, Shining Cuckoo, Sacred Kingfisher, Grey Warbler, korimako/Bellbird, tūī, pīwakawaka/New Zealand Fantail, Tomtit, tauhou/Silvereye, Welcome Swallow, New Zealand Pipit.

ed-billed Gull

Liz Light

Red-billed Gulls

Glenda Peake

## Banded Rail

*Gallirallus phillippensis*

Banded Rails are shy, margin-dwelling birds that live in some of the reed areas on the edges of Waimea Inlet. They are more common in the upper North Island and Great Barrier Island. There may be only 200 pairs left in the Nelson Marlborough area and none in the rest of the South Island. The birds live among rushes on the edges of estuaries, and feed on snails, crabs and worms that live on the intertidal margins. The Battle for the Banded Rail Project is aiming to increase the numbers of Banded Rails and other shy margin-dwelling birds. The project supports local communities in planting to restore degraded habitat and to protect the birds by pest control. It was established in 2014 and there are now more than 718 traps across the nine project sites and more than 3,000 pest animals have been caught. About 25,000 trees have been planted. Recent surveys, primarily of footprints on the mud, have located the Banded Rail in areas where it has not been seen for many years.

## Welcome Swallow

*Hirundo neoxena*

These attractive adroit, fast-flying, long-winged, fork-tailed little birds self-introduced from Australia in the 1950s, in Northland, and have gradually spread throughout the farming and built areas of the North Island and upper South Island. They do not live in forest or alpine areas and are often seen close to wetlands or the coast, and around lakes and ponds. They swoop and catch small insects above water as they fly, and drink water and scoop up insects on the surface of the water while flying. Adults have rusty-red markings on the forehead and neck, and a black eye-stripe. The back and upper wings are blue-black. The tail is dark with white spots towards the ends of the feathers, showing as a row of dots when the tail is spread in flight. The underparts are pale.

## Accommodation

Nelson is a city with a range of accommodation, including hotels, motels, camping grounds and backpacker hostels. If you would like a touch of estuary-side luxury, and birdwatching at all hours, Te Koi Lodge (tekoithelodge.com) at the end of Bronte Road is special.

# Kahurangi National Park

azetted in 1996, Kahurangi is New Zealand's newest ational park – and at 517,335ha it is the second largest atural protected area, almost filling out the north-west orner of the South Island. The area is rich in minerals nd timber, and protection came after a long, bitter battle etween conservationists and mining/logging lobbies. Its ue wealth arguably lies in its tracts of unspoilt wilderness nd extraordinarily diverse flora. There are forests of beech nd podocarp, nīkau palm groves on the wild west coast, and lpines on the tops. The park boasts more than half of New ealand's 2,400 native plant species, including over 80 per ent of its alpines (which have escaped the ravages of ice ages nd, later, goats).

irders are spoilt for options. The Oparara Basin, for instance, s a spectacular karst landscape of water-sculpted limestone nd marble, with tannin-stained rivers flowing through aves, arches and crops. As well as 18 of New Zealand's native irds, Kahurangi is home to ancient fossils (from rocks up to 00 million years old) and invertebrate giants, such as the Nelson cave spider and fist-sized, carnivorous *Powelliphanta* nails. North-west of the national park lies the Mangarākau wamp, where you may glimpse the Australasian Bittern or he mātātā/Fernbird.

There are some 570km of walking tracks – most famously he Heaphy, a multi-day tramp (or mountain bike) through ussock fields, wetlands, lush forest and coastal foothills, ffering the chance to encounter takahē and roaroa/Great Spotted Kiwi. If you aim to stay a few nights in Kahurangi, plan ahead (pre-booking for Heaphy Track huts is essential) and note that this is true wilderness: expect all kinds of weather, pack for all emergencies, and tell someone of your ntentions before you leave.

*The expansive beauty and mixed habitat of Kahurangi National Park*

## KEY FACTS

**Getting There**
The park may be accessed via roads out of Karamea (in the west), Murchison (south), Takaka (north) and Motueka (east). All four towns are linked by public transport, but check before you travel that roads are not closed due to snow, slips or flooding. Various local and national air carriers serve Nelson, Westport, Takaka and Karamea.

**Habitats**
Alpine herb fields, karst landscape, lakes and rivers, tussock downs, beech forest, podocarp forest, nīkau groves, wetlands.

**Best Time to Visit**
Kahurangi is good year-round for birds. Weather-wise, summer is usually fine and dry, yet busy, so you may prefer autumn. Winter is very wet and windy.

**Useful Websites**
www.nzbirds.com/birding/heaphy; www.doc.govt.nz/parks-and-recreation/places-to-go/nelson-tasman/places/kahurangi-national-park/things-to-do/tracks/cobb-valley-tracks; www.doc.govt.nz/globalassets/documents/parks-and-recreation/tracks-and-walks/nelson-marlborough/cobb-valley-mt-arthur-tableland-brochure.pdf; www.karameainfo.co.nz/oparara-basin; http://mangarakauswamp.com.

## Tracks and sites

**Heaphy Track** (4–6 days). This 78km (one-way) track is one of New Zealand's Great Walks. It may be tackled from either end: Kohaihai on the west coast, or Brown Hut in the Aorere Valley. To use the huts, buy a Heaphy pass from DOC. The Heaphy experience is too broad to cover here, but a highlight for birders is the section on the Gouland Downs – an open panorama of stunted shrubs, rust-brown tussocks and outcrops of eroded rock up to half a billion years old. Listen out for roaroa/Great Spotted Kiwi here, around Saxon Hut and at Perry Saddle. In 2018, 30 takahē were released on Gouland Downs in a bid to re-establish a local population; the first eggs were found in November that year – evidence of the first wild birds in a century to stake a foothold outside of the species' Fiordland refuge.

**Cobb Valley** (various walks; visit DOC online for details). Glacial action carved the big, U-shaped Cobb Valley, but the ice ages spared the diverse alpine flora on the tops. Here you might spot kea, the alpine parrot. Listen out for the rasping call of kākā, another parrot, or for the wheeze-whistle of whio/Blue Ducks, which haunt the fast-running streams. For access, take the Cobb Valley Road out of Upper Takaka.

**Oparara Basin** (various walks). Entering the Oparara Basin is like stepping back in time – or at least into a *Lord of the Rings* set. In this ancient karst landscape, erosion of limestone has shaped the rock over millions of years into moss- and fern-clad gorges, arches and fantastical caverns, where bones of Moa and Haast's Eagle have been found. Oparara is a sanctuary for the roaroa/Great Spotted Kiwi, and you may also spot whio/Blue Duck, kākā and weka. High tourist numbers are threatening the purity of this fragile habitat, so take special care: leave no rubbish, touch as little as possible, and use your torch sparingly in caves to avoid disturbing spiders and other secretive wildlife. Access to the basin is via McCallums Mill Road, which branches off the Karamea–Kohaihai Road about 4km north of Karamea (access is limited to vehicles less than 2.8m high).

**Mangarākau Swamp** Though not officially part of the park, this 350ha wetland is fed by the rivers of Kahurangi. Most of New Zealand's swamps have been drained, and while Mangarākau has lost its original forest cover of kahikatea and pukatea, it remains defiantly wet. While ownership is split between DOC and Native Forest Restoration Trust, a society – the Friends of Mangarākau – manages the area on a day-to-day basis (and offers on-site accommodation in its 15-bed lodge). Look out for orchids, native fish, and a range of birds including the Fernbird, pīpipi/Brown Creeper, Marsh Crake, Spotless Crake and Australasian Bittern. In recent years, researchers and volunteers have been recording the booming calls of the bittern, in a bid to calculate the size of the local population, which appears to number at least a dozen during the breeding season.

### Key Species

Roaroa/Great Spotted Kiwi, Blue Penguin, Black Shag, Pied Shag, Little Shag, Black Swan, Paradise Duck, whio/Blue Duck, Grey Duck, Scaup, White-faced Heron, Australasian Harrier, kārearea/New Zealand Falcon, weka, Pied Oystercatcher, Variable Oystercatcher, Spur-winged Plover, pūkeko, Banded Dotterel, Arctic Skua, Southern Black-backed Gull, Red-billed Gull, Black-billed Gull, Black-fronted Tern, White-fronted Tern, poaka/Pied Stilt, kererū/New Zealand Pigeon, kākā, kea, kākāriki/Yellow-crowned Parakeet, Long-tailed Cuckoo, Shining Cuckoo, ruru/Morepork, Sacred Kingfisher, Grey Warbler, Rock Wren, Rifleman, pīpipi/Brown Creeper, Tomtit, South Island Robin, Welcome Swallow, korimako/Bellbird, tūī, pīwakawaka/New Zealand Fantail, New Zealand Pipit, Fernbird, Marsh Crake, Spotless Crake, Australasian Bittern, tauhou/Silvereye.

*Brown Creeper*

Sid Marsh

## Roaroa/Great Spotted Kiwi
*Apteryx haastii*

The north-west corner of the upper South Island has probably always been the main stamping ground of the Great Spotted Kiwi, or roaroa. This is the second-largest of the five kiwi species, reaching a weight of 3kg in females and a little less in males, and it is reasonably numerous in the tussock grassland, scrub and forest, extending up to the subalpine zone. Great Spotted Kiwi can live for up to 50 years, but they breed slowly, rearing no more than one chick annually, and fall prey to introduced predators, especially stoats and dogs. Between 2010 and 2016, DOC and Friends of Flora transferred a total of 44 birds (captured from various localities) to a pest-protected area in the Flora Valley region in the north-east of Kahurangi National Park, to help boost genetic diversity in the local population. The first local chick hatched in 2012, and by 2018 Friends of Flora had solid evidence that their transplanted population was self-sustaining.

## New Zealand Rock Wren
*Xenicus gilviventris*

Visitors from Europe and the Americas take note: New Zealand's rock wren is unrelated to wrens elsewhere in the world. This pretty little yellow-flanked bird is endemic, and is one of just two survivors of an ancient stock of perching birds (the other is the Rifleman). Found across scattered alpine and subalpine parts of the South Island, the rock wren feeds mainly on insects and spiders, and nests in ground cavities. In this harsh environment, snow storms can dent wren populations, but a far greater threat comes from stoats and mice, which invade the nests. Studies by DOC in Kahurangi National Park during 2014–2017 show the terrible effect of these predators, but also the promise of effective control. In areas that had been treated with 1080 poison, nesting success stood at 58 per cent. In non-treated areas, it was just 13 per cent. While researchers continue to finesse protection for mainland populations, some birds at least have been removed to safer habitats – on, for instance, Secretary Island in Fiordland.

### Accommodation

Back-country huts (book ahead) on the walking tracks; full range of hotels, hostels and motels in the main local towns.

# Farewell Spit

Farewell Spit stretches for 30 km east of the top of the South Island

Imagine 12,000 Black Swans, 9,000 Australasian Gannets, 10,000 Red Knots and 12,000 Bar-tailed Godwits – these are some of the many bird species that can be seen on Farewell Spit from September to March. At the top of the South Island, Farewell Spit is a 30km sandspit with extensive tidal mudflats, extending 6km into Golden Bay at low tide. It is a bird sanctuary of international importance and 11,380ha of the spit are recognized under the RAMSAR Convention as being vital for shorebirds. This area is managed by DOC as a Nature Reserve and Shorebird Network Site, and is strictly protected. Apart from the first 4km at the base of the spit, visitors need to access the area through permitted tour organizations. Farewell Spit is particularly important for migratory shorebirds on the East Asian–Australasian Flyway. These birds nest in the Arctic, during the New Zealand winter and Arctic summer, but return to the spit for the summer months.

Eighty-seven bird species have been recorded at Farewell Spit, including, as mentioned above, about 12,000 Black Swans. There is a colony of 9,000 Australasian Gannets nesting here in September–March, and the spit is also home and breeding grounds for Caspian Terns, Southern Black-backed Gulls, Red-billed Gulls and Variable Oystercatchers. Besides these species, kororā/Little Penguins, herons, stilts and shags feed on the tidal flats.

## KEY FACTS

**Getting There**
This area is strictly protected. Apart from the first 4km at the base of the spit, visitors can only access the spit through permitted tour organizations. See details under Walks and tours overleaf.

**Facilities**
Farewell Spit Cafe is the northernmost in the South Island and there are spectacular views of the spit. Contact: www.whararikibeachholidaypark.co.nz.

**Best Time to Visit**
This important birding area has good birding in all seasons but thousands of migratory birds are present between October and early March.

## Walks and tours

Except for the 4km at the beginning of the spit, and there are quite good birding walks in this area, the spit can only be accessed by a licensed tour operator. Farewell Spit Eco Tours offers tours that are particularly suited to birders. Contact: www.farewellspit.com.

**Wader Watch Tours**, maximum four people, $135.00, duration two hours. This tour runs between September and April and the focus is on migrant waders. The journey is timed to the rising tide to observe the birds flying in to roost as the tide chases them across the intertidal plain. This is where the migratory Arctic waders come to feed and you will see Bar-tailed Godwits, Red Knots, Ruddy Turnstones and others. Timing depends on tides and weather conditions.

**Gannet Colony Tour**, $170.00, duration 6½ hours. There are about 9,000 gannets nesting between September and April. The gannet colony is 2km beyond the lighthouse and a 20-minute walk across the flat sand on the spit's end. You will see many other bird species as well. Take the bus back to the lighthouse keepers' cottage for refreshments. While there you can explore the reserve and soak up the serenity. The last stop is to climb one of the huge dunes for a look into Golden Bay and beyond. The Gannet Colony Tour is limited to 20 people per day so it is important to book in advance. Gannets usually choose rocky outcrops as breeding and nesting sites but the Farewell Spit colony is the only one in New Zealand that is right at sea level, affording excellent views of the parent birds caring for their young.

### Key Species

Blue Penguin, Australasian Gannet, Black Shag, Pied Shag, Little Shag, Black Swan, White-faced Heron, Reef Heron, kōtuku/White Heron, Cattle Egret, Royal Spoonbill, Australasian Harrier Hawk, Pied Oystercatcher, Variable Oystercatcher, Spur-winged Plover, New Zealand Dotterel, Banded Dotterel, Large Sand Dotterel, Golden Plover, Wrybill, Eastern Bar-tailed Godwit, Ruddy Turnstone, Red-necked Stint, Red Knot, Arctic Skua, Southern Black-backed Gull, Red-billed Gull, Caspian Tern, White-fronted Tern, poaka/Pied Stilt, Whimbrel, Sanderling, Pectoral Sandpiper, Sharp-tailed Sandpiper, Curlew Sandpiper.

Oscar Thomas

*Red Knots*

Oscar Thomas

## uddy Turnstone

*renaria interpres*

he Ruddy Turnstone is the third most umerous Arctic migrant wader species New Zealand, and of an estimated total opulations of 28,500 birds (and declining) bout 2,500 fly to New Zealand for the ummer. They live in coastal areas, on rocky r stony shores, sandy beaches or mudflats. hey arrive in New Zealand in August–lovember and depart northwards in March–lay, usually forming small flocks near the oast before departure. Of the migratory vaders they are thought to fly the furthest, esting in Siberia above 60 degrees north. fter leaving New Zealand they stop off and eed on the Korean Peninsula, before flying n to their Siberian breeding grounds. 'he Ruddy Turnstone is a stocky, blackbird-ized bird with short, powerful legs and eet, and a wedge-shaped bill with which it urns over large shells, stones and pieces of lriftwood while foraging for crustaceans. 3reeding adults have brightly patterned eddish, brown, white and tortoiseshell lumage on the back and wing-tops, and vhite underbelly and underwing.

## Banded Dotterel

*Charadrius bicinctus*

The Banded Dotterel is New Zealand's most common small plover and is found on seashores, estuaries and riverbeds. Its plumage changes seasonally but the birds are easily identified by their brown upperparts and chestnut band around the breast, and a black neck-band above this. This is most noticeable on the breeding plumage. Banded Dotterels breed on the braided rivers of the South Island, on rocky shores and river fans in the North Island, and in some subalpine areas, near water. Inland-breeding birds migrate to estuaries and other coastal wetlands the length of New Zealand after the breeding season, and may also travel to south-east Australia. Farewell Spit is one of the three most important wintering sites for the Banded Dotterel, and in winter there are some 800 birds in the area. Non-breeders may stay over the summer.

## Accommodation

Wharariki Beach Holiday Park is the closest accommodation. It has 30 tent sites, eight powered sites, four cabins and a backpackers' hostel that sleeps 18 people. Contact: www.whararikibeachholidaypark.co.nz. The Takaka/Collingwood region is especially beautiful and there is a variety of accommodation. Try Bookabach, Airbnb and Booking.com.

# Paparoa for Westland Petrels

There is only one nesting area for Westland Petrels, near Punakaiki on the west coast of the South Island. In the breeding season they can be seen at sunrise and sunset as they leave and return to their burrowed nests. The population, having declined radically since the advent of introduced predators, is stable at about 3,000 breeding pairs. The petrels are clumsy on land, and they waddle to their nests, often passing close to the viewing platform, or over it, if the nest happens to be on the other side. Usually birdwatchers get to see these birds at very close quarters and, later in the season, their chicks as well. The viewing shelter is positioned at a major landing and take-off site. Fledglings often use the viewing shelter seats for practice launches. During the New Zealand summer months, after the chick has fledged, it is believed that most of the population migrates to the sea off the western coast of South America. Westland Petrels are particularly susceptible to being killed during interaction with fishing. They follow fishing boats and are divers, so are susceptible to baited fish hooks.

Bruce Stuart-Menteath

*Forest habitat*

## KEY FACTS

### Getting There

3770 State Highway 6, Coast Road, Punakaiki, look for road signs 5km south of Punakaiki. Tel.: +64 3 731 1826, email: bruce@petrelcolonytours.co.nz. The site is on the property of Denise Howard and Bruce Stuart-Menteath, who usually personally conduct the tours. Boardwalks and wooden steps to the viewing platform have been constructed so that no burrows are damaged by people. Tours leave half an hour before dusk and, in some months (April–August), half an hour before dawn as well. Cost: $50 per adult, $10 per child. The tour takes about two hours.

### Best Time to Visit

The petrels are winter breeders and are present at their burrows from April until December when the chicks fledge. They spend the rest of the time at sea.

### Useful Website

Petrel Colony Tours covers the southern end of the nesting site. It provides the only way for birdwatchers to see these birds on the ground. There are approximately 130 breeding pairs in this area. For more information, see: www.petrelcolonytours.co.nz.

## Westland Petrel

*Procellaria westlandica*

The Westland Petrel, also known as the Westland Black Petrel, or tāiko, is endemic to New Zealand. The birds are large (length up to 48cm and weight up to 1.2kg), dark brown/black petrels with a cream bill. The legs and feet are black. They are burrowing birds and winter breeders, being present at their burrows from April until December when the chicks fledge. The birds form long-term monogamous pairs and lay just one large egg each year. Often, if not incubating their eggs or guarding their chicks, Westland Petrels fish during the day and return to their burrows at dusk, though they may stay away, hunting for food, for a week or more. They are elegant fliers but are clumsy on land and sometimes stumble and take a few moments to orientate themselves. The colony soon comes alive with squawks and cackles as more birds return and breeding pairs greet each other at their burrows. When the birds leave in the morning, to get airborne they climb on to a raised launch site, often old tree stumps, leaning trees or clifftops.

When they emerge from their burrow for the first time, fledglings may still be fully covered in grey downy plumage and spend most of their time sitting quietly near their burrow entrance. Others may have a mixture of down and adult plumage and explore their surroundings and exercise their wings. Older chicks display mature plumage and are very active, exercising their wings and practising launchings.

Bruce Stuart-Menteath

*Westland Petrel breeding grounds*

## Accommodation

Though this is a lightly populated area of the west coast, nearby Punakaiki Pancake Rocks are a well-known visitor destination and there are numerous accommodation options. This is a spectacular part of New Zealand with a wildly beautiful coastline. Coast Road Holiday Park, on the property of Petrel Colony Tours, has basic facilities for camper vans, tents and caravans.

# Ōkārito Lagoon and Ōkārito Kiwi Sanctuary

Located on the South Island's West Coast, sandwiched between the Southern Alps and the Tasman Sea, the Ōkārito Kiwi Sanctuary occupies about 11,000ha of lowland forest between the Waiho River and Ōkārito. Nearby is the Ōkārito Lagoon, with a small settlement occupying a sandspit at its southern end. The lagoon is relatively young, its reed-fringed waterways forged by a violent tsunami in the eighteenth century. At around 3,000ha it is New Zealand's largest unmodified coastal lagoon, home to around 76 native bird species, including the Royal Spoonbill and kōtuku/White Heron (which breeds nowhere else in New Zealand).

Māori had visited Ōkārito for centuries, but the 1860s gold rush transformed the sleepy stop into a bustling port, replete with hotels and grog shops. As gold fever ebbed in the 1880s, Ōkārito just as rapidly dwindled, and today around 30 souls call it home. Echoes of the past survive in the restored old buildings on the Strand, such as Donovan's Store (the village hall and live music venue). Today, it is ecotourism that draws visitors here, to see birds or explore the lagoon by boat or kayak.

The predator-free kiwi sanctuary is the sole mainland stronghold of the critically rare rowi/Ōkārito Brown Kiwi. Back in 1998, Forest & Bird and Kiwis for Kiwi had campaigned for the creation of 11 sanctuaries for kiwi nationwide; in 2000, the government approved five of these, including Ōkārito. A sole operator, Ōkārito Kiwi Tours, is licensed to take visitors into the sanctuary, and promises a remarkable 98 per cent chance of seeing kiwi. If you aim to do one of the walks, leave your dog behind.

## Tracks, walks and tours

See websites opposite for booking and other information.

**Ōkārito Kiwi Tours** are the sole operators licensed to guide visitors into the kiwi habitat. Night trips last around 2–3 hours; bring your own walking boots, insect repellent and patience! Pre-booking recommended.

**Ōkārito Kayaks** (formerly Ōkārito Nature Tours). Explore the waterways and teeming birdlife – either at your own pace with a freedom booking, or on a guided tour (depending on demand). Overnight booking also available.

### KEY FACTS

**Getting There**
Ōkārito is around 90 minutes' drive from either Haast (in the south) or Hokitika (in the north). At a point roughly halfway (15km) between Whataroa and Franz Josef on SH6, turn off on to Forks Ōkārito Road, then follow signs for 13km to Ōkārito. Come fully prepared, as you cannot buy petrol – or, indeed, much of anything – there.

**Facilities**
Parking, campsite. Some crafts for sale, and massage and yoga classes offered.

**Habitats**
Estuary, lagoon, sea cliffs, dense bush (rimu, rātā, silver pine, rainforest).

**Best Time to Visit**
Kiwi Tours run all year round. For kōtuku/White Heron and Royal Spoonbill, visit during the October–March breeding season.

**Useful Websites**
Ōkārito Kiwi Tours, http://okaritokiwitours.co.nz; White Heron Sanctuary Tours, http://whiteherontours.co.nz; Ōkārito Kayaks, www.okarito.co.nz; Ō kā rito Community Association, www.okarito.net; Ō kā rito Boat EcoTours, www.okaritoboattours.co.nz; Kiwis for Kiwi, www.kiwisforkiwi.org.

ārito Lagoon, with the Southern Alps in the background, on the west coast of the South Island

**kārito Boat EcoTours** This boat outfit offers range of tours, including the 80-minute rly Bird (best chance of seeing lots of rds), two-hour EcoTour (a comprehensive ploration of the lagoon and forest), and )-minute Wetlands Tour, which includes e kōtuku/White Heron breeding colony. our and accommodation packages also fered. Pre-booking recommended.

**kārito Wetland Walk** (1km, 20 minutes turn). This short, easy track leads from the hoolhouse car park through bush to the stuary and a viewpoint.

**kārito Trig Walk** (4.2km, 1½ hours return, asy). Begin on the Wetland Walk (above), ut after 10–15 minutes take a left to head phill to the trig viewpoint, where the pectacular vista takes in the Southern Alps, oast and lagoon.

**hree Mile Pack Track** (9.8km, 3½ hours eturn). Though the name is misleading, this onger walk is still easy going, but the full istance can be done only within an hour or so of low tide. From the Wetland Walk (above), follow signs over the Kohuamarua Bluff and through the forest to Three Mile Lagoon, where the track ends. Return the same way or (at low tide) via the beach.

## Key Species

Rowi/Ōkārito Kiwi, kororā/Little Penguin, tawaki/Fiordland Crested Penguin, Crested Grebe, Black Shag, Little Black Shag, Paradise Shelduck, Grey Duck, Grey Teal, Australasian Shoveler, Scaup, whio/Blue Duck, kōtuku/White Heron, Royal Spoonbill, Spur-winged Plover, Australasian Bittern, Australasian Gannet, Australasian Harrier, kārearea/New Zealand Falcon, weka, Southern Black-backed Gull, Red-billed Gull, Caspian Tern, White-fronted Tern, poaka/Pied Stilt, New Zealand Pigeon, kākā, kea, kākāriki/Yellow-crowned Parakeet, Long-tailed Cuckoo, Shining Cuckoo, ruru/Morepork, Sacred Kingfisher, Grey Warbler, Tomtit, South Island Robin, pīpipi/Brown Creeper, korimako/Bellbird, tūī, pīwakawaka/New Zealand Fantail, New Zealand Pipit, Rifleman, New Zealand Rock Wren, mātātā/Fernbird, tauhou/Silvereye.

Chrissy Wickes

## Rowi/Ōkārito Brown Kiwi

*Apteryx rowi*

At first glance the rowi looks much like any other brown kiwi, and in fact it only acquired full species status in 2003 – by which time it was in dire straits. Landmark research assembled by scientist John McLennan in 1996 revealed that almost all the surviving rowi were ageing adults (they can live to be 80 years old). There were almost no chicks being recruited into the population, due to predation by stoats and other predators. The numbers were shocking: only 5–18 per cent of hatched chicks were reaching adulthood.

The good news is that, while chicks are hapless, an adult kiwi can usually fend off mustelids with a good kick. And so in 1994, conservationists launched Operation Nest Egg (ONE). They take kiwi eggs from danger zones, hatch them in captivity and rear chicks to a 'safe' weight, before releasing them back into the wild in safer sites such as the Ōkārito Kiwi Sanctuary. It is chiefly due to ONE that rowi numbers have recovered from a low of 150–200 birds in the mid-1990s to approximately 500 today. All live in the Ōkārito sanctuary, bar a few that have been moved to offshore islands.

## White-faced Heron

*Egretta novaehollandiae*

The White-faced Heron is a tall bird with a blue-grey body and white face. It self-introduced from Australia in the 1940s, spread throughout New Zealand and is now abundant. It is most likely to be found in estuaries, swampy wetlands and even damp school playing fields after rain. The birds often roost in the tops of trees growin near water. They build a loose platform anc lay three to five eggs, which are incubated by both parents. Some breed in small colonies. In the cooler South Island they lay in October and incubation takes about 26 days. It is unusual for them to raise more than two chicks, though they often lay mor than two eggs. While foraging White-faced Herons walk slowly in still-water areas, with long, slow steps, and grab prey at spee with their sabre bills. During courtship and nesting, they prettily raise their plumes, an sometimes perform aerial displays.

## Accommodation

The Ōkārito campsite is run by DOC and Ōkārito Community Association (OCA); bring $1 coins for showers. There are a dozen or so cottages privately let; see OCA website for details. The old schoolhouse (built in 1901) has been restored into a 12-bunk hostel, run by DOC; for details and booking, visit www.doc.govt.nz/parks-and-recreation/places-to-go/west-coast/places/okarito-area/things-to-do/lodges/okarito-school-house.

# Naitangiroto Nature Reserve, kōtuku/ Nhite Heron Colony

ɔtuku, or the White Heron, *Ardea alba*, has legendary status New Zealand. It is a common bird in Asia but is rare here ith only one nesting colony, in the Waitangiroto Nature eserve near Ōkārito Lagoon, on the west coast of the South and. The colony was formed by the self-introduction of rds from Australia several hundred years ago. Australian rds are still occasionally blown over and add to the local ɔpulation. The herons are wetland birds and feed on small h, eels, frogs, shrimps, aquatic insects and mice. They have ong, sharp, dagger-like bill for grabbing prey.

his is the rarest of the herons in New Zealand. It has a ecial place in the hearts of New Zealanders and is on the ck of the $2 coin. It is loved for its elegance, beauty and ze. It stands up to 1m tall, and has bright white feathers d a yellow bill. During nesting, it grows a shower of long e feathers on its back. It is a graceful flier with a 150cm ingspan, and flies with its head pulled back with an shaped kink in its neck. There is a stable population of some 00 birds. They return to the colony in August and make their ests in the crowns of tree ferns and kōwhai trees. After the icks fledge, they spread to wetlands all over New Zealand. oyal Spoonbills and Little Shags also nest in the area.

**KEY FACTS**

Getting There
There is only one way to view this remote colony and that is through White Heron Sanctuary Tours, based in Whararoa on the West Coast, 30km north of Franz Josef Glacier. Contact: whiteherontours.co.nz; info@whiteherontours.co.nz. You are taken by jetboat and zoomed downriver to alight at a boardwalk. The walk, above swamp and through kahikatea rainforest, takes about 20 minutes. There are two hides from which you watch the birds across a small tidal creek. There are regular backpackers' buses travelling up and down the West Coast that stop at Whararoa.

Best Time to Visit
September–March.

*orest giants emerge far above the canopy at Waitangiroto Nature Reserve*

Liz Light

Liz Light

## Accommodation

White Heron Sanctuary Tours has a motel in the village and there are plenty of options for accommodation, from backpackers' facilities to luxury lodges, within 20km up or down the coast. See booking.com or Airbnb.

# Monro Beach

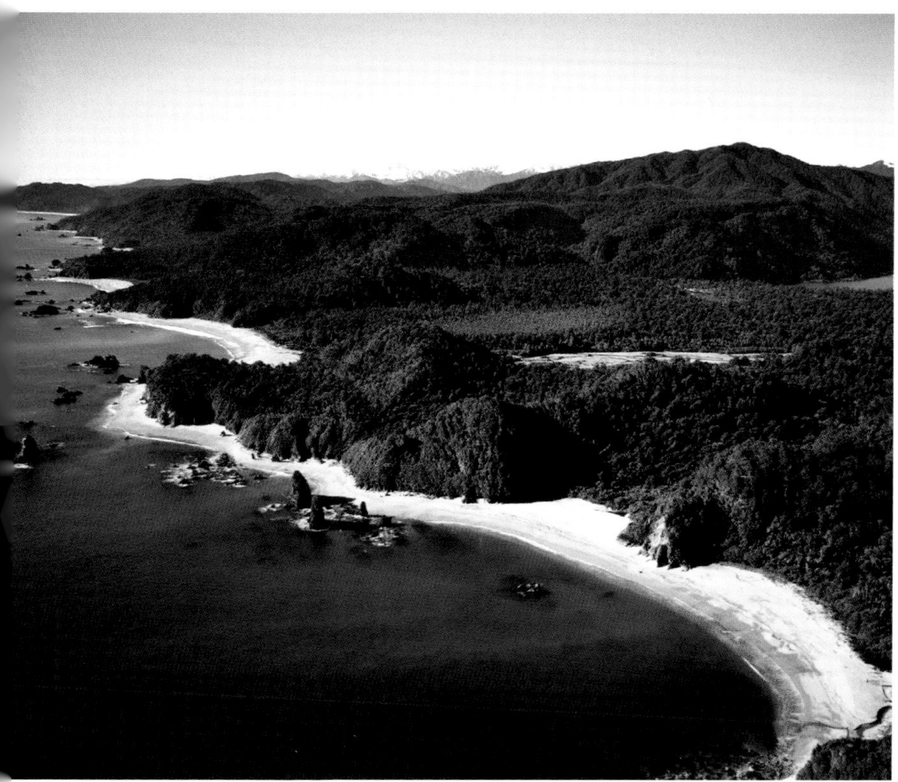

Monro Beach on the wild West Coast of the South Island

Fiordland Crested Penguins are only found in the bottom of the South Island and can be seen, in season, coming to shore in the early evening or leaving in the early morning at Monro Beach, near Lake Moeraki, 30km north of Haast on the West Coast, including Jackson Bay, or on the north-west circuit hiking track on Rakiura/Stewart Island. For more information, see www.tawaki-project.org/zbirdsonline.org.nz/species/fiordland-crested-penguin.

Fiordland Crested Penguins are highly susceptible to human disturbance when nesting. There is a concern that increased nature tourism in South Westland and Fiordland may disturb breeding birds and cause nests to fail. Stoats and dogs pose a serious risk to the bird colonies. Stoats prey on both chicks and sick or injured adults, while a single dog has the potential to wipe out an entire colony.

## KEY FACTS

**Best Time to Visit**
During the breeding season from July to November. The penguins can sometimes be seen during the moulting season from mid-January to early March.

## Tracks

A walking track leads from **Lake Moeraki** to the beach. Watch out for passerines and other forest birds along the path. Guided tours are conducted from the **Lake Moeraki Wilderness Lodge**. The best time of year to see the birds is during the breeding season in July–November. They can sometimes be seen during the moulting season from mid-January to early March. The penguins are very timid, so do not approach them, their nests or areas of beach where penguin tracks are common. At dusk stand back and watch them tentatively come out of the sea and waddle up the beach to the bush areas where they nest.

### Fiordland Crested Penguin/tawaki

*Eudyptes pachyrhynchus*

The Fiordland Crested Penguin, or tawaki, is one of the rarest of New Zealand's mainland penguins. It is an endemic bird and the current population is estimated to be 2,500 3,000 breeding pairs, but numbers have been in decline since the 1950s. Fiordland Crested Penguins are of medium height with dark blue-grey on the uppersides and white on the undersides. The special featur is the snazzy head markings, including a yellow eyebrow stripe that can stick up ferociously when a bird is agitated. The birds stand up to 60cm high, and they can be seen at dawn and dusk on some beaches and coves as they make their way from the sea to their nests in the bush. They nest in small colonies in late winter in loose nests in rainforest not far from the sea. Two eggs are laid in July and August – the first is smaller than the second, and they are laid three to six days apart. Often only one chic is reared, but in good years, two may be raised. The parents continue to feed their chicks until early December, when they go to the sea. They return for three weeks in February to moult.

Gerry McSweeney

*Fiordland Crested Penguins*

### Accommodation

Wilderness Lodge Lake Moeraki, is an easy 40-minute bush walk from Monro Beach. This place is special combining quality accommodation with exciting nature adventures and conservation. A naturalist can accompany you to Monro Beach. Local knowledge almost guarantees penguin sighting, plus seeing lots of passerine and other bird species along the way.
Ph +64 3 7500881 www.wildernesslodge.co.nz

# Makarora, Mt Aspiring National Park

Makarora braided river as it enters Lake Wanaka

### KEY FACTS

**Getting There**
On Highway 6 it is 50km from Hawea north to Makarora. The road is very good, the scenery is amazing and the journey takes about an hour. Heading south from Haast on Highway 6 it is 80km to Makarora and takes about an hour and a half. The road is usually in good condition, but in winter can be subject to slips and snow.

**Facilities**
Makarora is a sweet village strung out along the Makarora River valley. There are two cafes, a petrol station, an information centre and a pub/restaurant.

**Habitat**
Braided rivers (five species of shorebird nest in the braided rivers), alpine environments (alpine birds such as kea), and beech and podocarp forests (passerines and other forest birds).

**Best Time to Visit**
There is good birding in all seasons but more bird activity and song in spring.

Mt Aspiring National Park straddles the southern end of the Southern Alps and covers 3,500km$^2$. It is an area of supreme beauty with spectacular mountain scenery, alpine vegetation and valleys of beech and podocarp forest. It is part of Te Wāhipounamu – South West New Zealand World Heritage Area. World Heritage status acknowledges that this is among the world's most outstanding natural landscapes. The villages of Makarora, Haast and Glenorchy act as gateways to the park. Of these three areas Makarora has the most diverse birding and is a prime birding destination with a variety of habitats. The Makarora River, behind the village, is braided and is the late-spring nesting habitat for Wrybills, Black-billed Gulls, Banded Dotterels, Pied Oystercatchers and Black-fronted Terns. The forest next to town, especially in the early morning, is busy with kererū/New Zealand Pigeons, tūī, pīwakawaka/New Zealand Fantails, Tomtits, South Island Robins and kākāriki/Yellow-crowned Parakeets, and the occasional kākā or kea flies over the valley. There are other bird species in the higher forest areas and alpine regions that can be accessed by hiking or helicopter.

## Tracks and sites

**Makarora Tourist Centre** has brilliant birding from the tent or the cottage. No alarm clock is needed because the dawn chorus is a loud combination of tweeting, chiming and trilling. Kererū/New Zealand Pigeons swoop past, tūī squabble, pīwakawaka/New Zealand Fantails flit and Bellbirds sing. There is a 1km bush walk behind the accommodation, through giant podocarp forest, where you might see Tomtits, South Island Robins and kākāriki/Yellow-crowned Parakeets.

**Makarora River** runs down the valley behind the town and is braided for more than 15km before it fans out and reaches the headwaters of Lake Wanaka. The islands, braids and pebble banks are important nesting sites for Wrybills, Black-billed Gulls, Black-fronted Terns, Pied Oystercatchers and Banded Dotterels. Find out at the information centre where the public can access this river. Stay on the banks and use scopes and binoculars to spot birds. The nests and eggs are well disguised and easy to accidentally crush.

Higher up in the catchment of the contributing rivers lives the endangered whio/Blue Duck and higher still, in the subalpine areas, you can see New Zealand Rock Wrens. There are many tramping tracks in the valley and helicopters to take people to the alpine areas. Find out more at the Makarora Tourist Information Centre.

### Key Species

Paradise Duck, Grey Duck, whio/Blue Duck, Australasian Bittern, Australasian Harrier, kārearea/New Zealand Falcon, weka, Southern Black-backed Gull, Red-billed Gull, Caspian Tern, White-fronted Tern, Black-fronted Tern, Wrybill, poaka/Pied Stilt, kererū/New Zealand Pigeon, kākā, kea, kākāriki/Yellow-crowned Parakeets, Long-tailed Cuckoo, Shining Cuckoo, ruru/Morepork, Sacred Kingfisher, Grey Warbler, Tomtit, South Island Robin, korimako/Bellbird, tūī, pīwakawaka/New Zealand Fantail, New Zealand Pipit, Rifleman, New Zealand Rock Wren, Fernbird, tauhou/Silvereye.

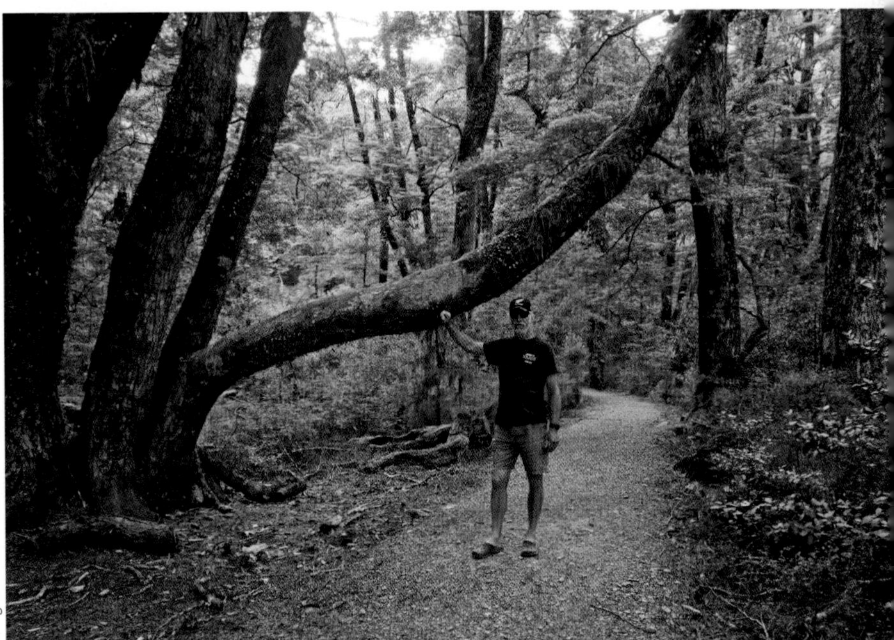
Liz Light

*Sam in the beech forest of Makarora*

## Black-fronted Tern

*Chlidonias albostriatus*

This is a small, attractive tern with distinctive breeding plumage. It has a dark grey back, pale grey underparts, a black cap, a white stripe coming from the bill below the cap and an orange beak and legs. Non-breeding adults have a speckled grey cap. The terns breed in spring and early summer on the braided rivers of the eastern South Island, including the Makarora River, and disperse to coastal areas after breeding, roosting and foraging in sheltered harbours, estuaries and lagoons, and offshore. They do not breed in the North Island but can be seen there in small numbers around estuaries in winter. They nest in small colonies on shingle bands, away from vegetation, in scrapes in sand or among river stones. The population of Black-fronted Terns, which may number 5,000–10,000 birds is declining for various reasons, including degradation and weed encroachment on braided rivers, and predation by mammalian predators and Black-backed Gulls. Trapping for pests on the Makarora River and control of Black-backed Gulls are giving these attractive birds a better chance. Adults return to the same river to breed each year.

Oscar Thomas

## Tūī

*Prosthemadera novaeseelandiae*

Tūī are found in all of the North Island and much of the South Island, and are an iconic songbird. The song combines bell-like notes with rasps, creaks and thumps, and in many areas this is the sound that wakes people in the morning. The tūī is dark – black, navy blue and brown – with a petrol-on-a-puddle iridescent sheen. Tūī have a tuft of white feathers on the throat, hence the sometimes name parson bird. They are aggressive defenders of their territory and their nests, and big enough (30cm long) to mean business. They are nectar feeders by preference but go for fruit in autumn. You will find them where trees, flaxes and shrubs are flowering or fruiting. Flaxes flower in Makarora in November and there are plenty of tūī around at that time.

### Accommodation

Makarora has bed and breakfasts, cottages, chalets, farmstays and campsites. Makarora Tourist Centre (www.makarora.co.nz) has a dozen dinky A-frame cottages, circa 1970, and camping areas. There is great birding from the cottages. Impressive early-morning bird chorus.

# Avon–Heathcote/Ihutai Estuary

Paul Corliss

*Manu-Bromley ponds at the estuary*

Christchurch, New Zealand's third biggest city, wraps around the Avon–Heathcote Estuary, some 880ha of intertidal flats, salt marshes, sewage oxidation ponds and wet grassland. The Avon and Heathcote Rivers merge in this estuary, as well as it being the catchment for lesser streams. A 4.5km sandspit separates it from the Pacific Ocean. This is a vast and varied habitat for wetland birds. In season, in autumn before the migratory birds leave, more than 30,000 birds are present here, from 38 bird species, with over 1,000 birds in 12 species. The area is immensely important to migratory birds and internationally recognized as part of the East Asian–Australasian Flyway network for migratory birds. Andrew Crossland is the Avon–Heathcote Estuary guru – he has been coming here and counting birds for 38 years. Those who would like more in-depth information can find it in a New Zealand Ornithological Society paper (see www.notornis.osnz.org.nz/system/files/Crossland%202013a.pdf).

## KEY FACTS

**Getting There**
There are nine reserves around the estuary. Each has different access points. Some are served by public transport (see www.metroinfo.co.nz). Christchurch is flat and a great city for cycling (see www.cyclingchristchurch.co.nz/bike-hire; www.naturalhigh.co.nz/biking/christchurch-bike-hire).

**Facilities**
The nine reserves link together, and each has different facilities. Each has well-maintained walking tracks.

**Best Time to Visit**
There are thousands of birds all year round but the best time, when there are the most birds, is January, February and March, before the migrant birds depart for the northern hemisphere and after the birds that nest inland have fledged and are congregating here.

## Tracks and sites

This is a huge area, and all nine reserves are linked around the estuary. They include Bexley Wetland, Lower Avon Saltmarshes, Raupo Bay Saltmarsh, Te Huingi Wildlife Reserve (the old sewage-treatment ponds), Linwood Paddocks, Charlesworth Reserve, Lower Heathcote Saltmarshes, McCormack's Bay Islands and South Shore Spit and foreshore. There are many walks and birdwatching sites in these areas, too many to list in this book. The best way to access information and maps about the numerous tracks is by Googling each of the above reserves.

Oscar Thomas

*Arctic Skua*

### Key Species

Black Swan, Paradise Shelduck, Grey Teal, Australasian Shovler, New Zealand Scaup, Little Shag, Black Shag, Pied Shag, Spotted Shag, kōtuku/White Heron, White-faced Heron, Royal Spoonbill, Swamp Harrier, pūkeko, Bar-tailed Godwit, Lesser Knot, Variable Oystercatcher, tōrea/South Island Pied Oystercatcher, South Island Pied Stilt, Banded Dotterel, Spur-winged Plover, Arctic Skua, Southern Black-backed Gull, Red-billed Gull, Black-billed Gull, Caspian Tern, Black-fronted Tern, White-fronted Tern, Sacred Kingfisher, Grey Warbler, pīwakawaka/New Zealand Fantail, tauhou/Silvereye, Welcome Swallow, New Zealand Pipit.

*The estuary at dusk*

Oscar Thomas

## New Zealand Scaup
*Aythya novaeseelandiae*

This is a dark, dear little duck with a profile rather like a bath toy. The male has a slightly iridescent green on its head and a bright yellow eye. The female is dark muted brown with brown eyes. Scaup are diving ducks and spend a lot of time underwater. This species does well on big lakes and has benefited from the creation of the large hydroelectric lakes in both the North and South Islands. There are thousands of the birds in the estuary, the Avon and Heathcote Rivers that feed into it and the Te Huingi Wildlife Reserve (the old sewage-treatment ponds). They congregate in reedbeds and under trees such as willows, and though they are good fliers they are not migratory.

## Red-billed Gull
*Larus novaehollandiae*

The Red-billed Gull is a common endemic coastal gull, and is often seen scavenging around fishing boats, rubbish bins, people with food on beaches and cafe gardens. It is usually a colonial nester and breeds on cliffs and rocky stacks, and attacks people who pass close by. It can be a noisy squawker and is not loved by fishermen or picnickers. It is an attractive white gull with grey wings that are trimmed with black, and a red bill, red legs and red around the eyes. The population is declining, possibly because of the closure of landfills and effluent discharges, and like all mainland birds it struggles in nesting sites without predator control around them. Predators include cats, rats, stoats and ferrets, but predator control is increasing.

### Accommodation

This wetland is in the middle of the country's third biggest city. There are hundreds of accommodation options, including backpackers' facilities, bed and breakfasts, motels, hotels and exclusive lodges.

# .ake Ellesmere/Waihora

:e Ellesmere

## KEY FACTS

### Getting There
The lake is approximately 50km south of Christchurch. Then the road extends some distance along the Kaitorete Spit, which gives visitors access to both the lake and the sea. It is a huge lake, divided into a number of separate reserves, and there are numerous access points. It is sensible to look at the map and brochure on the link below to decide where to focus your birdwatching.

### Best Time to Visit
Good for birding all year but the best time is in February and early March when there is an abundance of migratory birds, before they leave for the northern hemisphere.

### Useful Websites
For detailed information visit Waihora Ellesmere Trust website: www.wet.org.nz. There is a link to an excellent brochure and map on this site: www.wet.org.nz/wp-content/uploads/2010/09/WET-Lake-Access-Brochure1.pdf.

.is shallow brackish coastal lake is, at 20,000ha, the fifth gest lake in New Zealand. It is separated from the Pacific :ean by the Kaitorete Spit, an impressive 28km-long spit of .ooth river stones and shingle. The average depth is 1.4m .d it is rich in nutrients and has a variety of habitats that etland birds enjoy. DOC controls 35 per cent of the margins the lake, and there are numerous wildlife reserves around :hat are given extra protection. This is an internationally .portant birding area and New Zealand's single most :portant habitat for wetland birds. It has the greatest number species (167) and the highest bird numbers (approximately ),000) of any wetland in New Zealand. Grey Teal and Black vans are the most abundant birds, as well as Australasian .ovelers and Paradise Shelducks. Pied Stilts, Banded Dotterels .d Black-billed Gulls are also plentiful. Kaituna Lagoon, in e south-east of the lake, is good for waterfowl. The spit is eat for seabirds and the very end of the spit is good for shags, aterbirds and seabirds. This area is great for birding all year, ıt the migratory birds will be there in February before they ave for the northern hemisphere.

ıe vast brackish coastal lake is a haven for waterbirds and migratory shorebirds

## Tracks and sites

**Little River Rail Trail** A 13-kilometre section of this trail, from Motukarara to Birdlings Flat, runs alongside Lake Ellesmere, then Kaituna Lagoon. The old railway line was built on a raised embankment and this provides great viewing for birds. Both walkers and cyclers use this trail.

**Harts Creek Track** This is a 20-minute walk, along a riverbank to the lake, to a bird hide in a small wildlife reserve. From the Harts Creek car park follow the obvious track. There are boardwalks across the muddy spots so it is an easy walk for almost everyone. The final section of boardwalk can be covered in shallow water if the lake is high.

**Clarks Road Walk** From the end of Clarks Road walk to the gate at the western end of the paddock and follow a track that goes to the lake. There are hides about 1.5km from the car park. Waterproof footwear is recommended. You can return the same way or walk to Embankment Road to the west Jarvis Road to the east.

**Kayaks** provide a great way to quietly wat waterbirds. There are boat ramps at Wolfe Road, Lower Selwyn Huts and Lakeside Domain. There are places to launch at a number of other locations.

### Key Species

Black Swan, Paradise Shelduck, Grey Teal, Australasian Shoveler, New Zealand Scaup, Australasian Crested Grebe, Little Shag, Black Shag, Pied Shag, kōtuku/White Heron, White-faced Heron, Royal Spoonbill, Swamp Harrier, pūkeko, Marsh Crake, Spotless Crake, Australasian Coot, Bar-tailed Godwit, Lesser Knot, Sharp-tailed Sandpiper, Red-necked Stint, Ruddy Turnstone, Pacific Golden Plover, Banded Dotterel, Wrybill, Spur-winged Plover tōrea/South Island Pied Oystercatcher, South Island Pied Stilt, Arctic Skua, Southern Black-backed Gull, Red-billed Gull, Black-billed Gull Caspian Tern, Black-fronted Tern, White-fronted Tern, Sacred Kingfisher.

Liz Light

*Black-billed Gull*

## Tōrea/South Island Pied Oystercatcher

*Haematopus finschi*

South Island Pied Oystercatchers are handsome birds, with contrasting black-and-white plumage and long red bills. In spring and summer they breed inland on braided rivers, often near the base of the Southern Alps, and move to coastal estuaries for autumn and winter. They are plentiful (possibly over 100,000 birds) and survive predation by aggressively guarding their territory and their chicks.

Oscar Thomas

## .ustralasian Shoveler

*.nas rhynchotis*

ustralasian Shovelers are filter-feeding 'aterbirds with shovel-shaped bills that are road and flat at the end. They filter feed n the surfaces of lakes, rivers and brackish 'aters. They are sexually dimorphic, with ıe male being brightly coloured (grey head with a vertical stripe, white-and-grey speckled breast, rusty flanks) and the female being a much less glamorous dull mottled brown. They are resident in the lake all year round but in bigger numbers in November–June, peaking at more than 1,000 in January. They move around New Zealand a lot but usually return to their traditional breeding areas.

## Accommodation

Lakeside Domain has camping facilities right on the lake's edge. Nearby, the sweet village of Little River has Little River Camp Ground, which also has cottages, and Okuti Gardens has self-contained accommodation.

# Arthur's Pass National Park

Liz Light

*On the road towards Arthur's Pass*

Gazetted in 1929, Arthur's Pass was the South Island's first national park, and its origins lay in the search for a manageable east–west route across the alps that would give access to the new-found Hokitika goldfields. In the summer of 1864, Arthur Dobson and his brother Edward set out to find it. Informed by local Māori, who had long used the pass as a pounami (greenstone) trading route, they took their horses up the Waimakariri River, then over and down to the confluence of the Rolleston and Bealey Rivers. Arthur gave a somewhat reluctant thumbs-up to the precipitous pass that now bears his name, and after some hefty engineering, the route opened in 1866. By 1923 there was a much easier and faster rail service, via the 8.5km Otira Tunnel, and visitors flooded in, prompting conservationists to call for protection in the form of a national park. But you can still take a car up the scenic SH73, which winds heart-stoppingly over the tops. TV producer Gus Roxburgh described Arthur's Pass as 'a park with a split personality'. Certainly, with a kea's-eye view, you would see how the spine of mountains creates two weather systems. The prevailing nor'westers shed their rain – up to 7,000mm of it annually – on the western flanks, where the deep valleys are cloaked in dense rainforest. In the much drier east are beech forests flanking broad braided rivers. These rivers have their resident birds, but also many migrants that visit in warmer months to breed. In the beech forests you may spot the mohua/Yellowhead and kākāriki/Yellow-crowned Parakeet, while the rainforests harbour diverse small birds, such as Tomtits and Riflemen. If you are

## KEY FACTS

**Getting There**
Road access is via SH73, a spectacular drive from either direction. From the west, leave SH6 at Kumara Junction, midway between Greymouth and Hokitika. From here, you are looking at about 80km (one hour) beside the Taramakau and Otira Rivers and over the impressive Otira Viaduct, to Arthur's Pass. From the east, it is 154km (two hours) from Christchurch, out along the Waimakariri via Darfield and Springfield, to the national park. As alternatives to driving, take the Tranzalpine Express train, which stops at Arthur's Pass, or a coach from Greymouth, Hokitika or Christchurch. Fill up on fuel and food at the nearest main centre before pressing on to Arthur's Pass village, where goods are pricier.

**Habitats**
Rainforest, beech forest, braided rivers, steep gorges, scree slopes, alpine plateau, rocky peaks.

**Best Time to Visit**
There is more bird activity and song in spring, otherwise there is good birding in summer and autumn. Winter can be harsh and the area is often snow-covered.

**Useful Websites**
www.arthurspass.com; www.doc.govt.nz/parks-and-recreation/places-to-go/canterbury/places/arthurs-pass-national-park/bird-watching; www.arthurspass.com/pdf/arthurs-pass-brochure.pdf.

ky, you might see or hear the roaroa/ at Spotted Kiwi at night, along with the u/Morepork. The high spine itself, with scree slopes and rugged peaks and alpine adows, is the haunt of the feisty kea and ust little New Zealand Rock Wren.

## acks and sites

**thur's Pass Walking Track** (6.8km, 2 urs 40 minutes return). This easy track ds from Arthur's Pass village to the bson Memorial at Arthur's Pass summit. e longest option is from the Punchbowl park. Walk up to the Bridal Veil Falls kout, then cross the creek and wind through the alpine meadows to Jacks t. Cross the highway to Bealey Chasm park and head up through the beech est, keeping your eyes peeled for bird e. When you intersect with the Dobson ture Walk, switch to this track to head to the summit, where a lookout gives a ew of the memorial. Alternatively, re-cross e highway to the Temple Basin car park. turn the same way to the village. No dogs.

**Devil's Punchbowl Walking Track** (2km, one hour return). Beginning at the Punchbowl car park, cross the Bealey River and Devil's Punchbowl Creek, then head up through the beech forest (looking out for Tomtits) to a viewing platform at the base of the waterfall, which is spectacular after heavy rain. The lucky may spot kea here. Return the same way. No dogs.

**Temple Basin Track** (3km, three hours return). This easy-to-medium grunt begins at Temple Basin skifield car park beside SH73, 5km north of Arthur's Pass village. Wind up to the alpine meadows and tussocks for great views and a chance of spotting the New Zealand Rock Wren.

**Hawdon Valley** From the village, drive east on SH73 for 24km, then turn left on to Mount White Road, crossing the river. Fork left again and continue to the road end, where there is parking at the Hawdon Valley Campsite. From here, tracks lead into the valley, or up the ridge via the Woolshed Hill Track. The forests here are alive with birds

*braided river on the east side of Arthur's Pass*

Liz Light

*A semi-frozen waterfall near Arthur's Pass township*

### Key Species

Black Shag, Little Shag, kōtuku/White-faced Heron, Paradise Shelduck, Canada Goose, Mallard, Australasian Harrier, pūkeko, tōrea/ South Island Pied Oystercatcher, Pied Stilt, turiwhatu/ Banded Dotterel, Wrybill, Spur-winged Plover, karoro/Southern Black-backed Gull, tarāpuka/ Black-billed Gull, tara/Black-fronted Tern, Paradise Shelduck, whio/Blue Duck, Rock Pigeon, Shining Cuckoo, Sacred Kingfisher, Welcome Swallow, New Zealand Pipit, Grey Warbler, pīwakawaka/South Island Fantail, tauhou/Silvereye, New Zealand Rock Wren, toutouwai/South Island Robin, mohua/ Yellowhead, roaroa/Great Spotted Kiwi, kea, kākā, weka, korimako/Bellbird.

– mohua/Yellowhead, Tomtit, Rifleman, pīpipi/Brown Creeper and kākāriki. If you fancy overnighting, the Hawdon Hut is 9k (three hours) from the road end.

**Waimakariri River** There are many good birding spots along the upper reaches of the Waimakariri, including Bealey Bridge, Hawdon Flat and Turkey Flat (approached via Crow Valley Track). As well as year-rou residents, migrants come to breed from September to January. Look for threatened species such as the Wrybill, tara/Black-fronted Tern and tarāpuka/ Black-billed Gull, as well as the Paradise Shelduck, kākā, karoro/Southern Black-backed Gull, Spur-winged Plover, turiwhatu/Banded Dotterel, Canada Goose and tōrea/South Island Pied Oystercatcher. To avoid damaging nest sites, avoid walking on bare gravel, and heed any 'keep off!' warnings from the birds themselves.

**Otira Valley Track** (1–3 hours). This challenging track offers the chance to see kea, weka, New Zealand Rock Wren and whio/Blue Duck. Begin at the car park to the left of SH73 about 6km north of the village. Follow the marked track across the moraine and through the scrub to the Otira River footbridge (one hour). From here on, the route across the scree to the head of the valley is unmarked, and suitable only for experienced and fit trampers; there is also risk of avalanches in winter and early summer. Return the same way.

### Kea
*Nestor notabilis*

The world's only alpine parrot, the kea lives in the high country of the South Island, including Arthur's Pass, where you may see it raiding rubbish bins, clattering upon hut roofs, or wheeling high over the valleys, uttering its raucous *keeeeaaa* cry. A large, bronze-olive bird with red nape and underwings (in both sexes), it has a long, hooked bill, which it uses mostly to gather berries and grubs. Along the way, it plays a vital role in dispersing the seeds of alpine plants.

Oscar Thomas

a are super intelligent, highly resourceful d quick to learn new tricks, but their bold riosity is too often fatal. In the early years ey were heavily persecuted, particularly their habit of pecking into the backs of eep to access fat, which is largely why ey are now restricted to remote high rain. Today, though no longer legally lled, farmers still take revenge on a few rds; kea also annoy visitors by ripping to car wipers and door seals and stealing mp gear, and they gorge on junk food left ing around, which encourages the birds hang around settlements and pester ople. Troublingly, kea enjoy the taste of ad and will tear the toxic metal from old of flashings or car-wheel rims, leading to rious sickness and death. Today there are me 5,000 kea remaining, and the species listed as nationally threatened. Much search is under way, by DOC and groups ch as the Kea Conservation Trust, into how ese delightful and unique birds can coexist ith the humans that encroach on their oorstep.

### South Island Robin

*Petroica australis*

The South Island Robin is a small bird with long legs, black eyes and short, fine bill. Both sexes are dark grey above and pale grey below; the two colours are clearly demarcated on the male's chest, more blurrily so on the female's. The male is the more avid songster, too, especially in the breeding season, when he proclaims his territory and advertises for a mate. Favouring mature forest and scrub, robins feed on invertebrates of all types from worms to wētā, and also on ripe fruits. Breeding right through from July to December, a pair of South Island Robins can raise four clutches of 2–4 young in a year. Introduced predators are a threat to the female robin particularly, as it is she who looks after the nest and fledglings, resulting in a skewed sex ratio today, but pest control is an effective countermeasure, as is translocating birds to predator-free zones. It can be relatively easy to approach robins closely, as they tend to be confiding birds, especially on well-used walking trails.

## Accommodation

Arthur's Pass village offers the full range – hotel, motel, bed and breakfasts, backpackers' facilities – but places are limited and you should book well in advance if visiting in summer. There is also accommodation further afield, for example at Bealey, Otira and Cora Lynn. More details available at www.arthurspass.com.

# Tasman River and Black Stilt

Department of Conservation

*The Tasman River valley with Aoraki/Mt Cook in the background*

For birders the Tasman River is all about the rare and beautiful kakī/Black Stilt *Himantopus novaezelandiae*. It has the dubious status as being one of the most endangered birds in New Zealand. The story of the kakī is another back-from-the-brink-of-extinction story that New Zealand bird lovers are only too familiar with. This attractive endemic stilt, whose adult plumage and bill are totally black and legs are bright red, was, by 1981, down to 23 birds. Its natural habitat, in the South Island's braided riverbeds, side streams, swamps and ponds and sometimes lake margins, was severely curtailed by the draining of wetlands for farming and the flooding of rivers and swamps for hydroelectric dams. It is also particularly susceptible to predation by feral cats, stoats, weasels, ferrets, rats and hedgehogs. Additionally, human disturbance (vehicles and walkers) can crush the birds' eggs and scare them from their nests. Now it is found in the wild in the Tasman River Valley before it enters Lake Pukaki.

## KEY FACTS

### Getting There

The road from Pukaki town to Mount Cook goes alongside Lake Pukaki and then alongside the delta area. Do not drive off the road on the riverbed. To find out more go to the Twizel Information Centre, Market Place, Twizel, Tel.: +64 3 435 3124; twizelinfo@gmail.com.

### Best Time to Visit

Spring and summer, which is nesting season. Most birds leave after the chicks fledge and for the winter months.

ıce 1981 there has been an intensive ptive breeding programme, in both ıristchurch and Twizel, managed by ͻC with enormous help from the Isaac ›nservation and Wildlife Trust. Kakī eggs e artificially incubated and the young ıicks are raised in captivity. At 3–9 months d they are released into the wild. Rearing em in captivity significantly increases eir chances of survival by preventing edation when they are most vulnerable. ıis also reduces the chance of nesting ults being taken by predators. There are ›w 132 birds in the wild, and intensive edator trapping in the Tasman River area is creasing the chances of the birds breeding their natural habitat. Most other riverbed rds leave the area for the winter and igrate to the coast. but kakī tend to stay.

ıkī are carnivorous, eating insect larvae, olluscs, worms and spiders. They often rm lifelong pairs and due to low numbers, hen they cannot find a mate they may breed with Pied Stilts. Hybridization is not such an issue now the numbers have increased and with intensive management. They lay 3–4 eggs in nests between stones. In the Tasman River area, where there is intensive predator control, there are 15 breeding pairs, and the chances of chicks reaching breeding age has increased to about 50 per cent.

The Tasman River flows south for 25km through the wide, flat-bottomed Tasman Valley, in the Southern Alps, and into the northern end of the glacial Lake Pukaki. The best place to see these birds is in the area where the delta of the river joins Lake Pukaki. Besides kakī, Wrybill, Black-fronted Terns, Black-billed Gulls, Banded Dotterels and Pied Oystercatchers nest in the riverbed. Take great care when walking in this area, especially during the nesting season in spring and early summer. The nests of all these birds are between river stones. Watch where you put your feet.

*lack Stilt*

## Accommodation

This beautiful lake and mountain area is popular with visitors and locals alike, and there is all sorts of accommodation around Mt Cook Village, Tekapo Village, Pukaki and Twizel, all relatively close to the Tasman River.

# Oamaru Blue Penguin Colony

The Oamaru Blue Penguin Colony is a fully fledged tourist operation, but without its restrictions, controls and the trapping of predators, the penguins would be doomed. The penguins have lived on this site for hundreds of years, way before the town of Oamaru expanded into their territory, and the only way to protect them is to restrict visitors to specific areas and times. It is a tourism operation with a strong research and conservation focus, ensuring that the penguins are protected and that the population achieves long-term stability. A tour during the day allows visitors to peek into the penguins' nests, stroll through their habitat and learn about their lives. In the evening, visitors watch the nightly arrival of penguins from their day's fishing at sea. Reviews on Trip Advisor are mixed. Some complain about the cost, the other tourists and the controls. Others say it was a fantastic experience and well worth the money. I, personally, enjoyed it and learnt a lot. See www.penguins.co.nz.

### KEY FACTS

**Getting There**
The colony is in Oamaru, a sweet heritage and farming town, three hours' drive south of Christchurch Airport and 1½ hours north of Dunedin city. The colony is five minutes from the centre of Oamaru town, and is one of the easiest places to see the Little Penguins. See website for the directions map. Price: day entry from $18, evening viewing from $32, combo day and evening viewing from $45. Email: bookings@penguins.co.nz; Tel.: 03 433 1195. Take warm clothes. Evening viewing lasts for a couple of hours as the penguins arrive home. It can get chilly.

**Facilities**
Interpretation centre with lots of information about penguins, shop and toilet facilities.

**Best Time to Visit**
There are penguins here all year but fewer in March.

Glenda Peake

*Little Penguin*

### Little Penguin
*Eudyptula minor*

Little Penguins, also known as Blue Penguins, Little Blue Penguins and kororā, are the smallest penguin species in the world, standing at 30cm. They normally live to around eight years, and can raise up to four chicks per breeding season (in two different clutches). Little Penguins swim up to 50km per day, during the breeding season, to feed their chicks. Incubation of the eggs takes up to 36 days and the chicks fledge 7–8 weeks after hatching. The birds have a range of calls to recognize each other, to claim and defend territory, and to communicate before coming ashore in groups (rafts), which can number more than 100. They normally forage as individuals but meet up to come ashore. When foraging at sea, they stay under the water for only 20–30 seconds at a time and dive an average of 800 times a day. For two weeks between January and March penguins stay ashore continuously during their annual moult, but otherwise they are only active on land at night.

Oscar Thomas

## Accommodation

Besides the penguins, Oamaru is a popular destination as it is a quirky heritage town with a population of 15,000. There are all sorts of accommodation options.

# Royal Albatross Centre, Pukekura, Taiaroa Head

Pukekura, the Royal Albatross Centre

*Royal Albatross chick and Taiaroa Head Lighthouse*

Pukekura, the Royal Albatross Centre, is at Taiaroa Head, on the tip of Otago Peninsula, and is just 30 minutes' drive from Dunedin city. This is the only place on the mainland where these majestic birds nest. Since the first albatross egg was laid at Pukekura, in 1920, the colony has slowly grown to 50 nesting pairs, with more juveniles returning as adults, ready to mate, every year. The survival of these birds has increased with the management of DOC, predator control and control of visitors.

The Royal Albatross Centre is an eco-tourism destination, with an information centre, cafe and shop. Visitors are managed as these birds are too endangered and rare to have people wandering willy-nilly around their nesting site. Maximum group numbers are 18. It is the only easily accessible place in the world where people can see these extraordinary birds and you can often view them quite close up from two underground hides.

The Royal Albatross Centre also has a Little Blue Penguin colony of some 180 pairs (see page 191) and here at Pilots Beach just below Taiaroa Head, you can see the birds in their natural habitat emerging from the sea and scurrying up to their cliff-face burrows. After a day at sea, the penguins congregate in groups known as rafts not far offshore, where they are often heard vocalizing – usually uttering short, loud

## KEY FACTS

**Getting There**
It is on the end of Otago Peninsula, 30km from the city of Dunedin. Follow Portobello Road and then Harington Point Road. For those without transport there are daily pick-ups from Dunedin city. Visit the I-centre in central Dunedin. There are guided walks for a maximum of 18 people.

**Facilities**
Interpretation centre, toilets, cafe, souvenir/educational material shop.

**Fees**
The albatross tour is $52 for an adult, less for children, and there are family group deals. It takes about an hour, includes a short and informative film, and visits to two hides where you can see the birds quite close up as well as, on a breezy day, see these massive birds soaring overhead.

**Best Time to Visit**
Albatrosses nest biennially and the breeding process takes just about a year, from mating in October to fledging the following September. There are usually birds there for the entire year but likely to be fewer between mid-September and mid-October.

**Useful Website**
www.albatross.org.nz.

uawks. At dusk they come ashore and ake their way to their nests, where they ed their chicks or roost. Viewing is at dusk, e time of which changes during the season. iewing is managed and costs $35 per adult, ss for children, and there are concession ices for families. There are penguins here l year but very few in March.

esides the albatross and penguins, a umber of other bird species nest at Taiaroa ead, including some 4,000 Red-billed ulls and colonies of Spotted Shag, the rare ewart Island Shag and Royal Spoonbills.

## Iorthern Royal Albatross

*iomedea sanfordi*

orthern Royal Albatrosses are endangered, omprising some 7,000 breeding pairs. Most nest on Chatham and other Southern Ocean islands. This is the world's largest bird, weighing 8kg or more, with a 3m wingspan. The birds nest biennially and in the year between breeding they fly around the Southern Ocean alone, before returning to their nesting site, in September, to reacquaint with their mate. They meet and after an elaborate courtship with lots of squawking, cackling and cracking of beaks, re-establish their relationship, one that lasts for life. They lay only one egg. The first eggs of the colony are usually laid in November. They hatch in late January and the chicks fledge 240 days later in September. Juvenile birds stay away, fishing and flying alone at sea, before they return to Taiaroa Head when they are 4–5 years old, but they do not usually first breed until they about 7–8 years old. Eighty per cent of the fledged birds return, the rest are lost, probably through human fishing activities.

## Accommodation

Camping, motorhomes and cabins. Portobello Tourist Park is in a sheltered valley behind Portobello village. It has 11 powered sites. From $18.00 per night. Tel.: 03-476 1006, email: www.portobellopark.co.nz. Dunedin is a city of approximately 130,000. There is the full range of accommodation, from backpackers' lodges and camping grounds through to posh hotels.

# Otago Peninsula and Yellow-eyed Penguin

Liz Light

*Otago Peninsula*

## KEY FACTS

**Useful Websites**
For more information about the hoiho, see www.yellow-eyedpenguin.org.nz. To see hoiho on the Otago Peninsula, visit Elm Wildlife Tours, www.elmwildlifetours co.nz. The tours leave Dunedin every afternoon, take 6½ hours in summer and 5½ hours in winter, and comprise small groups of up to 12 people, at a cost of $112 per person. Although the cost is substantial, this is likely to be the only opportunity you get to see this unique and endangered penguin. A tour also covers a lot of other birds and wildlife, and takes you to isolated beaches that would otherwise be illegal to access. I have been on this tour and it is excellent. Take a snack and warm clothes. A fitness level is required as there is quite a lot of walking.

**Best Time to Visit**
Mating begins in July and fledging happens in March. The best time to visit is September to March. Weather in Dunedin can be staunch in the winter.

A few private beaches on the Otago Peninsula are mainland breeding colonies for the hoiho, the Yellow-eyed Penguin. These birds are vulnerable and endangered, and unescorted viewing of them is not possible. The beaches they nest on are on private land and have conservation status. Elm Wildlife Tours has access to these sites and is involved with conservation activities for hoiho. Besides the highlight of these tours, watching hoiho nesting from hides, they provide an opportunity to see and photograph Hooker's Sea Lions, a breeding colony of New Zealand Fur Seals, Little Blue Penguins and numerous other estuarine and marine birds.

Liz Light

*Yellow-eyed Penguins return to their nests from the sea*

### oiho/Yellow-eyed Penguin

*legadyptes antipodes*

oiho are handsome birds. They are tall and olid with a pale yellow band of feathers assing around and behind their bright ellow eyes. The back is dark blue-grey, and ne chest, belly and underside of the flippers re white. They have pink feet and a pale reamy bill.

loiho are completely endearing, in an ngainly way. They are sleek and swift vhile swimming in the waves, but when ney come ashore their elegance turns into vaddling clumsiness, as if they are drunk nd trying not to stumble. They do not dally n the beach, where they could be predated y seals or sea lions. When they are a little vay up the hill and out of danger, they stop to rest, preen their feathers and survey the landscape before their next climb.

Hoiho live in coastal forests, scrub or dense flax, and their nest sites are carefully selected to be shaded, sheltered and well concealed. Pairs meet at the nest in June–July, lay two eggs in mid-September, and both sexes incubate the eggs, which hatch in 40–50 days. The chicks fledge in March–April, depending on the laying time, and the adult birds forage for fish 2–25km offshore. They are nationally endangered and, sadly, their numbers are declining, both on the mainland and in their main breeding grounds in Campbell and Auckland Islands. There are estimated to be only 255 breeding pairs on the mainland and the decline is attributed primarily to marine impacts such as climate change and overfishing.

## Accommodation

Dunedin, a city of approximately 130,000 people, is on the doorstep of Otago Peninsula. There is the full range of accommodation from backpackers' lodges and camping grounds through to posh hotels.

# Fiordland National Park

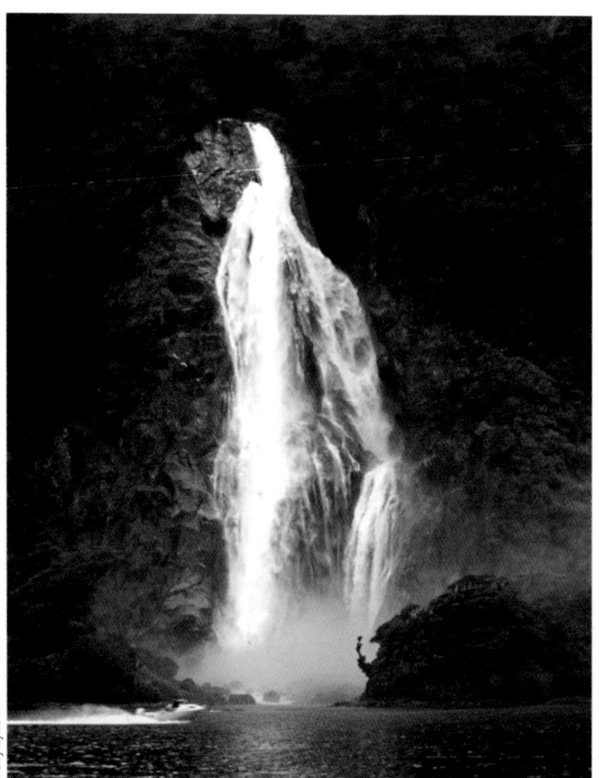

Kelly Lynch

*Spectacular scenery and hundreds of waterfalls, Fiordland National Park*

At 12,607km$^2$, Fiordland National Park, on the south-west corner of the South Island, is New Zealand's largest national park. Along with Mt Aspiring National Park it makes up Te Wāhipounamu World Heritage Site, an area of outstanding natural beauty. Much of the park is inaccessible by road, some is accessible by hiking tracks, including two of the Great Walks, and some is accessible by boat. There are nine bird sanctuary islands that are off-limits to visitors. Geographically, the park includes dramatic fiords, pristine lakes, deep, U-shaped valleys, icy mountains and dense forest. Beech-podocarp forests dominate, along with extensive areas of temperate rainforest with hundreds of different species of fern and moss. The park is administered by DOC. Many rare and endangered birds are found here. Birdwatching is terrific, particularly for those who have the time to do one of the Great Walks.

## KEY FACTS

### Getting There

The national park is huge and there is not a lot of road access. Highway 94 from Te Anau to Milford is 120km in length. There are good day walks and birding sites from various places along this road. See tracks and sites. The best birding is from the Great Walks, but they each take four days and need to be booked well in advance.

### Facilities

There are good-quality huts on the Great Walks and various other walking paths in the park. There are no supermarkets beyond Te Anau, so if you are camping or hiking stock up on supplies here.

### Habitat

This huge park has it all. Beech and podocarp forest, ferny rainforest, lakes, rivers, glacial valleys, sea sounds, swampland, alpine valleys and beaches.

### Best Time to Visit

Late spring to early autumn. The birding is good year round though the weather can be challenging in winter.

### Useful contacts

Fiordland National Park is managed by DOC, which has an excellent visitor centre on Lakefront Drive, in Te Anau. Tel.: +64 3 249 7924, email: fiordlandvc@doc.govt.nz.

## 'acks and sites

**ie Milford Track**, one of New Zealand's eat Walks, extends for 54km from the rthern end of Lake Te Anau, to Sandfly int near Milford Sound. It takes four days, n only be walked one way and has a limit the number of people that can be on the ack at one time. Bookings are essential and ust be made through the DOC website. ices vary according to season. There is a xury way to do this walk, where clients ly carry their day packs. It takes five days ıd four nights, and costs from NZ$2,130.00. e www.ultimatehikes.co.nz. The walk has ccellent birdwatching along the way.

ay 1. Tūī, korimako/Bellbirds, Tomtits, wakawaka/New Zealand Fantails and, on ie lake, Paradise Shelducks and shags.

ay 2. Watch out for pīwakawaka/New ealand Fantails, Tomtits, robins, tūī, orimako/Bellbirds, Riflemen, Grey Warblers, a and kākāriki/Yellow-crowned Parakeets.

Day 3. See many of the previous birds plus, at the higher altitudes of McKinnon Pass, kea are usually about.

Day 4. This passes lakes and streams, and adds a variety of ducks to the bird count, including the rare whio/Blue Duck.

**The Kepler Track** is a 67km, moderate walking track that takes four or five days to complete. It traverses lake edges, beech forest, alpine mountain tops and a U-shaped glacial valley. Great birding and highlights include the South Island Brown Kiwi near Iris Burn Hut and Mt Luxmore, whio/Blue Duck near Iris Burn Waterfall, kārearea/ New Zealand Falcons near Moturau Hut, kea at Mt Luxmore, kākāriki/Yellow-crowned Parakeets, robins and pīpipi/ Brown Creeper in numerous areas and New Zealand Pipit in the open areas on the ridge tops. Bookings to walk this track must be made through DOC. Prices vary according to season.

*Ailford Sound, Fiordland National Park*

**Other bird sites** besides the great walks. This is a link to a DOC brochure on different day walks: www.doc.govt.nz/globalassets/documents/parks-and-recreation/tracks-and-walks/southland/fiordland-day-walks.pdf.

1. Knobs Flat, see www.knobsflat.co.nz. Knobs Flat is on the Te Anau to Milford road. There is good cabin and camping accommodation as well as guided birding by an expert local. Mohua/Yellowheads, whio/Blue Ducks, New Zealand Rock Wrens, kārearea/New Zealand Falcons and Black-fronted Terns are some of the less common bird species that can be seen around Knobs Flat or close to it. Alongside these birds the kāka, kākāriki, Rifleman, Tomtit, South Island Robin, korimako/Bellbird, tūī, pīpipi/Brown Creeper, ruru/Morepork and many others are common to the area. Guided birding: half day $60, full day $120.

2. Homer Tunnel entrance. This is the place to see kea, the world's only alpine parrot. But beware, they are known to steal passports, remove the rubber from car windscreen wipers and do other naughty things with vehicles. Also, kārearea/New Zealand Falcons can be seen on the rocky slopes nearby, and sometimes the New Zealand Rock Wren.

3. Mirror Lakes. This walk is good for waterbirds as well as some forest birds along the path.

## Key Species

Tokoeka/South Island Brown Kiwi, tawaki/Fiordland Crested Penguin, kororā/Little Penguin, Little Shag, Back Shag, Pied Shag, Paradise Shelduck, pāteke/Brown Teal, Grey Duck, whio/Blue Duck, kōtuku/White Heron, White-faced Heron, Royal Spoonbill, Australasian Bittern, Australasian Harrier, kārearea/New Zealand Falcon, weka, Variable Oystercatcher, Southern Black-backed Gull, Red-billed Gull, White-fronted Tern, Pied Stilt, takahē, kererū/New Zealand Pigeon, kākā, kea, kākāriki/Yellow-crowned Parakeet, Long-tailed Cuckoo, Shining Cuckoo, ruru/Morepork, Sacred Kingfisher, Grey Warbler, Tomtit, New Zealand Rock Wren, Rifleman, South Island Robin, korimako/Bellbird, tūī, pīwakawaka/New Zealand Fantail, New Zealand Pipit, mohua/Yellowhead, pīpipi/Brown Creeper, Fernbird, tauhou/Silvereye.

Glenda Peake

*Takahe*

Paul Hersey

*The Kepler Track, Fiordland National Park*

Oscar Thomas

### ākāpō

*rigops habroptila*

kāpō are very large (2kg) flightless night
rrots. The arrival of humans and their
ammalian predators almost caused the
mise of the species, and only 50 remained
the 1980s. In 2018 the population was
0. It increased only after all the birds were
ansferred to three predator-free forested
lands in the Fiordland National Park,
d with intensive human intervention
pecially during chick rearing. Sadly, they
e now at risk again from a disastrous
ngal infection. Each kākāpō is named by
e Kākāpō Recovery Programme, and is
acked through a radio transmitter.

heir breeding behaviour is erratic: they may
nly breed once every four years, laying up
three eggs. The males boom loudly – this
the world's loudest bird – and the sound
avels over several kilometres, but this acts
a dinner gong for predators. Kākāpō are
orgeously green with mottled yellow and
ecks of black, and their danger-evasion
trategy is to freeze, which gives introduced
nammalian predators a walk-up invitation.
lew Zealanders do not seem to mind
he millions of dollars spent ensuring the
xistence of these odd but endearing parrots.

### Great Crested Grebe

*Podiceps cristatus*

The Great Crested Grebe is indigenous to New Zealand but is also found in Australia. It is a handsome bird with two parallel black crests on its head and chestnut frills on its cheeks. It is unusual (and cute) in that it carries its young on its back. It is only found in the large South Island lakes and, after years of declining in numbers, predator trapping and artificial nesting platforms have allowed the numbers to increase. The easiest place to see the grebes is on the shoreline of Lake Te Anau, often nesting along the 4km between the DOC office and the outlet of Lake Te Anau. They build their nests on submerged branches like an anchored pontoon.

## Accommodation

See DOC PDF for camping sites in the park. Te Anau, a medium-sized town, has a variety of accommodation on offer. Knobs Flat accommodation includes six studios at different prices depending on size and number of people staying. Camping and camper vans are $20 per person. See www.knobsflat.co.nz. Blue Thistle Cottages, (www.bluethistlecottages.com), near Te Anau, has cottages set among forest and good birding on the property.

# Waituna Lagoon

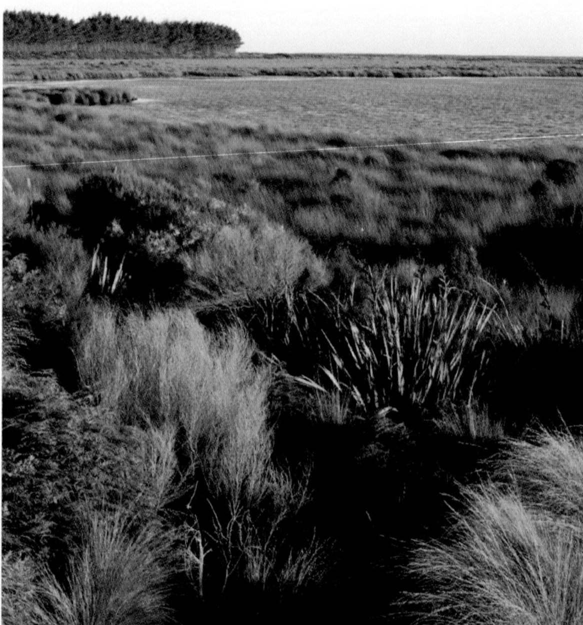
Environment Southland

*Waituna Wetlands*

## KEY FACTS

**Getting There**
Awarua Bay, the big eastern inlet off Bluff Harbour, is about 20km south-east of Invercargill. Head towards Tiwai, but before crossing the bridge at Awarua Bay, turn left and follow the shore road till you find parking. Be prepared for knee-deep wading. Waituna Lagoon, 40km from Invercargill, can be accessed from Moffat, Waghorn and Waituna Lagoon roads.

**Habitat**
Mudflats, peat moorlands, gravel bars, reedbeds, swamp margins, estuarine waterways.

**Best Time to Visit**
Good for birding all year but the best time coincides with the migratory birds who arrive in October and leave in February and early March. This area is in the very south of New Zealand and the winter weather can be wet, cold and very windy.

**Useful Websites**
https://southlandnz.com/the-catlins/natural-attraction/waituna-lagoon-awarua-wetlands; www.doc.govt.nz/parks-and-recreation/places-to-go/southland/places/awarua-waituna-wetlands; www.waituna.org.nz.

At the foot of the South Island, to the south-east of Invercargill, lies one of the most important remaining wetland areas in New Zealand. The 1,350ha Waituna Lagoon is part of a larger wetland complex, including Awarua Bay, which covers some 20,000ha of freshwater swamps, peatlands and estuaries. It includes the 3,500ha Waituna Wetland Scientific Reserve, which in 1976 became the first New Zealand wetland to be registered on the international RAMSAR Wetland Convention. Protection was expanded in 2008 to cover the entire complex.

Katrina Robertson

*Waituna Gravel Pit*

llectively, the Awarua–Waituna wetland osystem is a vital refuge for more than 80 d species, several threatened native fish d unusual plant communities – including oalpines – that thrive in the cool, damp mate. In summer it is home to more siting waders than almost any other site the country: these include the Sanderling, arp-tailed Sandpiper and Greenshank. ar-round residents include the Marsh ake and Fernbird.

e brackish lagoon itself, fed by three main eeks, was a traditional harvesting area for gāi Tahu (its name means 'water of eels'), d it is still popular among anglers, walkers d duck shooters, but it is also susceptible toxic algal blooms. Given the wetland's ique and valuable biodiversity, multiple cal and national bodies are working partnership to preserve it within the ider Arawai Kākāriki wetland restoration ogramme.

e lagoon's sea outlet is opened manually, ually once a year, so its hydrology and linity are highly variable. This is a wild region where access to birdwatching areas can be tricky underfoot and requires tidal awareness, and it is wise to prepare for all weather conditions.

## Tracks and sites

**Waituna Lagoon Boardwalk** (10 minutes return). This easy 250m boardwalk on the northeastern shore, accessed via Waghorn Road, connects to a birdwatching shelter overlooking the lagoon. Toilet and picnic table at car park.

**Wetland Loop Track** (60–90 minutes). From a point 50m along the boardwalk (above) you can divert to this easy (wheelchair-friendly) 5km track around wetland bog and mānuka thickets, with good viewing over the lagoon.

**Waituna Quarry Restoration Track** (five minutes return). The Waituna Landcare Group is converting this former gravel quarry into a wetland. Access to the viewing point and interpretation board is via a 300m gravel track leading from a car park on the corner of Waituna Lagoon Road and Hanson Road.

*Diggers at Walker's Bay*

**Awarua, head of the bay** (2½ hours). Drive east along Awarua Bay Road to the parking area at Muddy Creek, then walk around the head of the bay to access the bird-feeding and roosting areas. Be prepared for muddy wading! Birdwatching is best at high tide (60–90 minutes after high tide at Bluff).

**Beach walk** (4–5 hours return). From Waituna Lagoon Road you can walk as much or as little as you like of the white-sand beach, from the eastern end to the lagoon outlet. Do not drive on to the gravel or you will become bogged down.

## Key Species

White-faced Heron, tōrea/South Island Pied Oystercatcher, Variable Oystercatcher, Australasian Pied Stilt, New Zealand Dotterel, Banded Dotterel, Pacific Golden Plover, Spur-winged Plover, Ruddy Turnstone, Red Knot, Sanderling, Curlew Sandpiper, Sharp-tailed Sandpiper, Bar-tailed Godwit, Black Swan, Grey Duck, Grey Teal, New Zealand Shoveler, tītī/Sooty Shearwater, Black Shag, Little Shag, Stewart Island Shag, Marsh Crake, Caspian Tern, Black-fronted Tern, White-fronted Tern, Fernbird, Southern Black-backed Gull, Red-billed Gull.

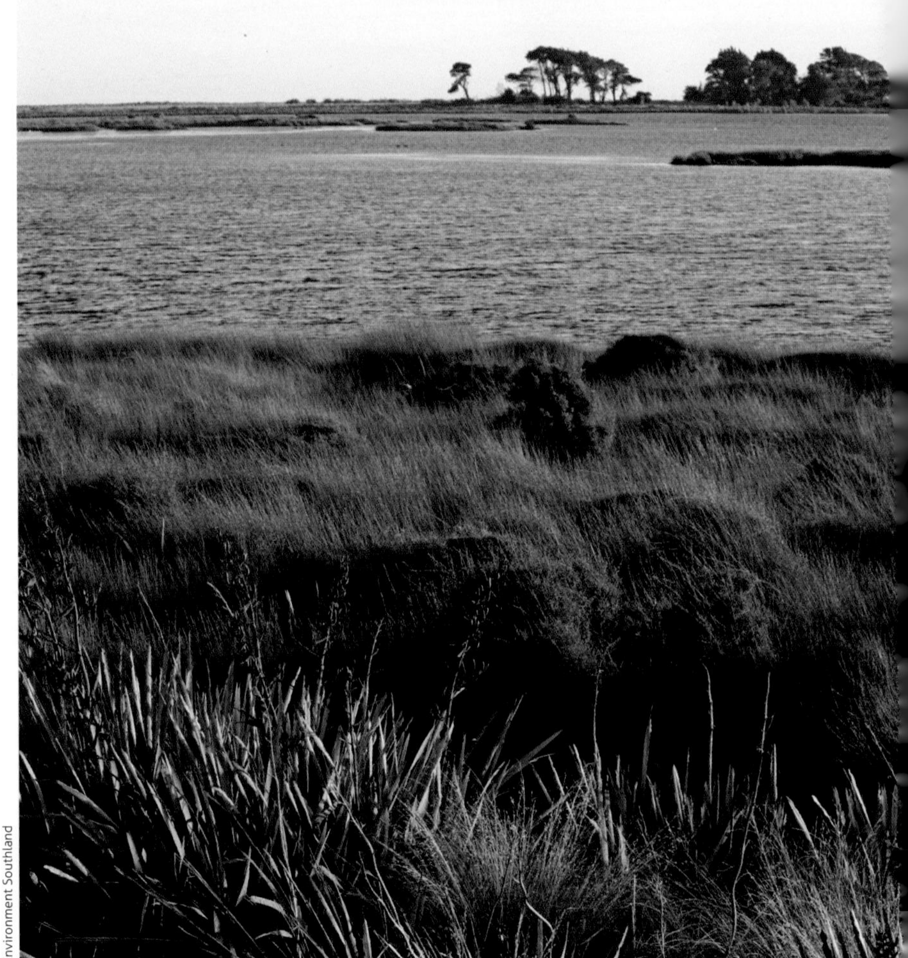

Environment Southland

*Wetlands, a vital refuge for more than 80 bird species*

Oscar Thomas

## ıarsh Crake

*ɔrzana pusilla affinis*

ɔu will be lucky to spot the secretive little arsh Crake, which skulks in thick wetland ·getation and is usually only heard at ght. But it is beautiful, with rich, flecked .estnut upper plumage, barred flanks ıd long olive-green legs. A member of e rail family, this New Zealand bird is an ıdemic subspecies of Baillon's Crake, which widespread around the Old World. The ıarsh Crake favours undisturbed reedbeds ıd is still reasonably widespread around ew Zealand, but the wholesale dwindling wetlands poses a threat to its survival, ıd even today it remains little studied. he Awaroa–Waituna complex is one of its ey strongholds, along with the Ashburton .kes/Ō Tū Wharekai in Canterbury and ıe coastal inlets around the top of the ɔuth. Breeding is in September–December; airs make a cup-shaped nest in marginal egetation, and the female lays 4–7 olive-rown eggs, which hatch in 2½–3 weeks.

## Sharp-tailed Sandpiper

*Calidris acuminata*

It takes a true expert to distinguish the 16 species of *Calidris* sandpiper that visit New Zealand: after all, most are a subtle variation on a theme, being speckled brown above and off-white below. The Sharp-tailed Sandpiper is one of four that visit regularly between September and April, having made a jaw-droppingly long migration from their northern Siberian breeding grounds. Sharp-tailed Sandpipers tend to settle in tidal harbours anywhere around New Zealand, as well as in the inland braided rivers of Canterbury; look for a reddish cap, fine white eye-ring and pale arched eyebrow. The bird's name derives from its pointed tail feathers, though they are not easy to assess from a distance. Today these birds are very rare visitors, with maybe only a couple of dozen appearing each year. Alterations to their stopover sites in the Yellow Sea are probably to blame.

## Accommodation

The nearest town, Invercargill, has plenty of accommodation and is within easy driving distance. If you are accessing the harbour and lagoon by boat, Bluff is closer; there is also a boat ramp at the end of Waituna Lagoon Road.

# Rakiura/Stewart Island

Kelly Lynch

*Rakiura/Stewart Island boasts bush, beaches and good birding*

Rakiura/Stewart Island is New Zealand's third largest island. It comprises 1,746km$^2$ and lies 34km south of the South Island. It is mostly covered in podocarp rainforest and is one of the best places to see both terrestrial and pelagic birds. The Māori name for the island, Rakiura, means 'glowing skies', referring to gorgeous sunsets and the famous Aurora Australis, the southern lights, which can often be seen here in the winter months. The population is just over 400, and most people live near the main town, Oban, in Halfmoon Bay. Lucky for the birds, the island is free of introduced animal pests such as stoats, ferrets, mice and goats. It does, however, have wild cats, rats and possums, which damage the habitat and birdlife. It is surrounded by 170 smaller islands and rocky islets, some of which are totally predator free.

Rakiura National Park takes up 85 per cent of Rakiura/Stewart Island, some 157,000ha. Trapping in areas of the park for cats, rats and possums has been helpful to the bird population, as has intensive trapping by the people who live in and around Halfmoon Bay. Rakiura/Stewart Island has around 25,000 kiwi, the tokoeka, a species unique to Rakiura/Stewart Island. Tokoeka make this one of the easiest places to see kiwi in the wild. It is a pelagic/seabird hotspot, with more than 20 species of seabird breeding on the island or islets; other seabirds from the subantarctic islands, and ocean-going species, gather and are supported by the fecund ocean around the island.

## KEY FACTS

**Getting There**
It is a 20-minute flight from Invercargill Airport and there are three flights a day (www.stewartislandflights.co.nz), or a one-hour daily trip on the ferry from Bluff.

**Getting around**
There are 28km of roads, mostly sealed, and you can hire bikes, motorbikes and cars on the island. The weather is notoriously changeable so always contact the visitor centre for the latest information and conditions before going on long walks. It is colder than most of the rest of New Zealand so take warm clothes, even in summer.

**Habitats**
Pristine rainforest, wetlands swamps, harbours and beaches.

**Facilities**
Boat, car and bike hire, a visitor centre, a pub, a cafe or two, a grocery store and other small shops.

**Best Time to Visit**
Summer and early autumn. The winter months are often really tough on Rakiura/Stewart Island with howling winds and sleety rain. The tracks can be muddy. Summer is best.

**Useful Websites**
www.stewartislandexperience.co.nz.

## ·acks and tours

·sit www.stewartisland.co.nz/walks. ·is is a pelagic/seabird hotspot with ·re than 20 seabird species breeding the island or islets. Other seabirds ·m the subantarctic islands, and ocean- ·ing species, can also be seen. There are ·merous boat owners who specialize taking clients on seabird-watching ·rs. Both Rakiura Charters (www. ·kiuracharters.co.nz) and Aurora Charters ·ww.auroracharters.co.nz) do full-day and ·lf-day pelagic birdwatching tours around ·kiura/Stewart Island. Ruggedy Range ·ilderness Adventure (www.ruggedyrange. ·m) is another recommended company.

·ere is a good chance of seeing hoiho/ ·llow-eyed Penguins and tawaki/Fiordland ·ested Penguins on the **beaches** at dusk on the Rakiura Track. This is a moderate three-day, 36km track, and is one of New Zealand's Great Walks. Blue Penguins can be seen around the wharf at Oban at dusk and dawn, and also on the beaches on the Rakiura Track.

**Ulva's Guided Walks** (www.ulva.co.nz) This company has four different guided walks for birdwatchers, including one to Ulva Island. They vary in length and price, and include the cost of boat travel to and from the walking tracks (see also Ulva Island, page 208).

There is good birding around **Oban township and Halfmoon Bay** because of habitat restoration and trapping for predators. Look out for passerines (korimako/Bellbird, tūī, Tomtit, pīwakawaka/New Zealand

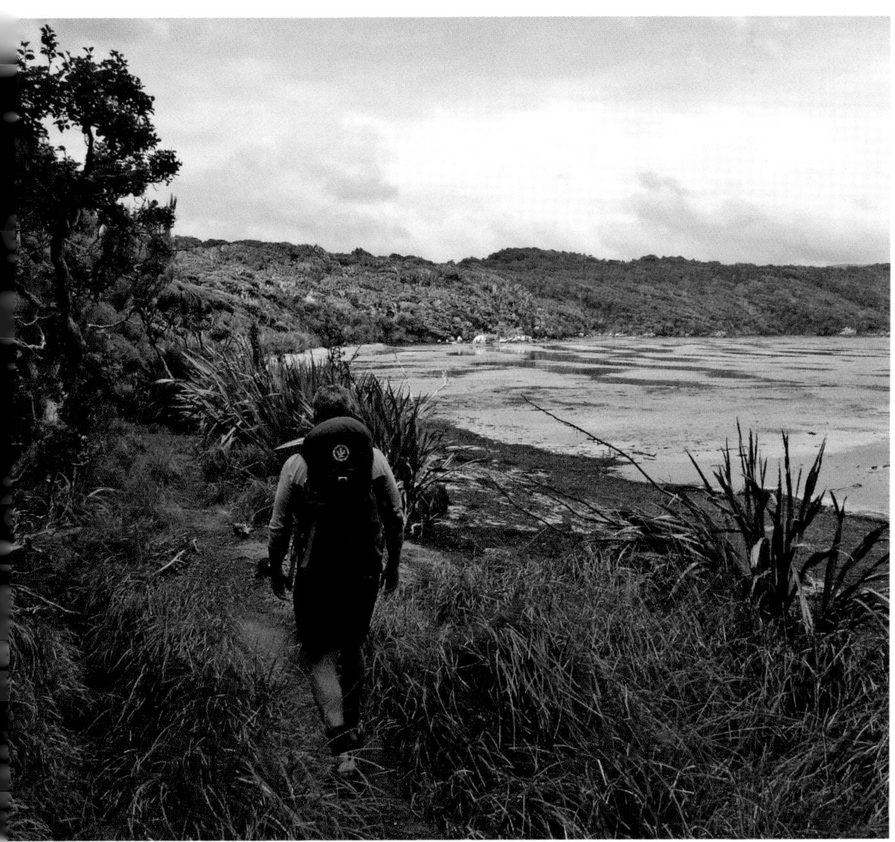

*·he Rakiura Track on Rakiura/Stewart Island; good penguin spotting in season*

## Key Species

**Seabirds:** hoiho/Yellow-eyed Penguin, Blue Penguin, tawaki/Fiordland Crested Penguin, Wandering Albatross, Northern Royal Albatross, Southern Royal Albatross, New Zealand Black-browed Albatross, Buller's Albatross, White-capped (Shy) Albatross, Salvin's Albatross, Light-mantled Sooty Albatross, Northern Giant Petrel, Antarctic Fulmar, Snares Cape Pigeon, White-headed Petrel, Mottled Petrel, Cook's Petrel, Broad-billed Prion, Fairy Prion, White-chinned Petrel, tītī/Sooty Shearwater, Grey-backed Storm Petrel, White-faced Storm Petrel, Black-bellied Storm Petrel, Southern Diving Petrel, Australasian Gannet.
**Shorebirds and forest birds:** Stewart Island Kiwi, Black Shag, Pied Shag, Little Shag, Stewart Island Shag, Spotted Shag, Blue Shag, White-faced Heron, kōtuku/White Heron, Reef Heron, Cattle Egret, Australasian Bittern, Black Swan, Paradise Shelduck, Grey Duck, New Zealand Shoveler, Australasian Harrier Hawk, kārearea/New Zealand Falcon, Banded Rail, Stewart Island Weka, tōrea/South Island Pied Oystercatcher, Variable Oystercatcher, Spur-winged Plover, Banded Dotterel, Southern New Zealand Dotterel, Eastern Bar-tailed Godwit, Ruddy Turnstone, Southern Skua, Southern Black-backed Gull, Red-billed Gull, Black-billed Gull, Black-fronted Tern, Caspian Tern, Antarctic Tern, White-fronted Tern, kererū/New Zealand Pigeon, kākāpō, South Island Kākā, kākāriki/Red-crowned and Yellow-crowned Parakeets, Shining Cuckoo, Long-tailed Cuckoo, ruru/Morepork, Sacred Kingfisher, South Island Rifleman, Welcome Swallow, New Zealand Pipit, pīwakawaka/New Zealand Fantail, Stewart Island Robin, Fernbird, pīpipi/Brown Creeper, mohua/Yellowhead, Grey Warbler, korimako/Bellbird, tūī, tauhou/Silvereye, tīeke/South Island Saddleback.
With thanks to Ron Tindal who compiled this list.

Liz Light

*Yellow-eyed Penguin*

Fantail, Grey Warbler and pīpipi/Brown Creeper). Stewart Island Weka also thrive in this area. The DOC has an excellent day walks brochure that can be picked up at the visitor centre or downloaded (www.doc.govt.nz/globalassets/documents/parks-and recreation/tracks-and-walks/southland/stewart-island-rakiura-short-walks-brochure.pdf). Try Ackers Point lighthouse walk at dusk, in late summer, to see kororā/Little Penguins and tītī/Sooty Shearwaters returning from fishing.

Spotting kiwi. A large population lives around the dunes and tussock grassland of **Mason Bay**. Seeing the birds involves a 40-minute water-taxi trip, then a four-hour walk. It is best to go on an organized tour, and take two days and enjoy one night in a back-country hut (www.ruggedyrange.com). Ocean Beach is also quite good for kiwi spotting and not so strenuous. It is a 35-minute boat trip from Halfmoon Bay and two hours easy walking. September–May (www.stewartisland.co.nz/organisations/tours/wild-kiwi-encounter).

## tī/Sooty Shearwater

*ıffinus griseus*

ıe Sooty Shearwater, also known as the uttonbird, is a common coastal seabird in e southern half of the South Island and is so seen around the coast in the rest of New aland. It is a large (40–46cm long) dark earwater with long, narrow wings, long, ender bill and a short tail. The upperparts e sooty coloured and the underparts are ghtly greyer with a paler area on the outer ea of the underwings. The bill is black, as e the feet, which tuck away in flight.

oty Shearwaters breed annually and turn to the same colony, often with the me partner, each year. After breeding ey migrate to the North Pacific Ocean – 92 er cent of breeders survive from one year the next and they can live for up to 20 ears. The young can be harvested around akiura/Stewart Island by descendants of akiura/Stewart Island Māori. The harvest eason is in April and May, when the chicks e in their burrows or emerging from them. hey purportedly taste like fishy, fatty, salty utton. Despite harvesting by humans, the sk of predation on mainland colonies and ıe interaction with the fishing industry, hich kills hundreds of birds each year, ıere are millions of these hardy birds with rge populations around the Snares Islands nd Rakiura/Stewart Island.

Oscar Thomas

## Stewart Island/Foveaux Shag

*Leucocarbo chalconotus*

Previously thought to be one species, the Stewart Island Shag has recently been declared to be two species using DNA and morphometric analysis; the Foveaux Shag (living along the Foveaux Strait and Rakiura/Stewart Island) and the Otago Shag (living around the Otago Peninsula).

Large (68cm long and weighing up to 2.5kg), the pink-footed Foveaux Shag forages in groups that tend to fly close to the water. They have been known to forage as far as 10km from land. Foveaux Shags are colonial nesters and build pedestal-type nests of leaves, sticks and twigs cemented together by guano, and lay two or three eggs. They thrive on predator-free islets and can often be seen around the wharf at Halfmoon Bay. They are endangered by fishing, especially set netting, when they dive and get entangled in the mesh.

### Accommodation

Backpackers' facilities, one hotel, some very nice homestays and good tramping huts on the Rakiura National Park tracks.

# Ulva Island

Ulva Island, in pristine Paterson Inlet, off Rakiura/Stewart Island, comprises only 267ha but is one of New Zealand's best places for birding. It is named after an island in Scotland's Inner Hebrides, from which one of the early settlers came. The island, which is easily accessible by water taxi from Rakiura/Stewart Island, has a relatively unmodified habitat of mature podocarp forest. It is a bird sanctuary, which is open to the public, and is managed by the DOC. It has been free of all predators since 1997, when rats were removed, and is now a safe haven for birds, which are thriving. Rare and endangered forest species here include the Stewart Island Robin, tīeke/South Island Saddleback and Stewart Island Kiwi. The Rifleman and mohua/Yellowhead have returned to the island and now breed here. Besides these, you will probably see korimako/Bellbirds, tūī, kākāriki/Yellow-crowned Parakeets, pīpipi/Brown Creepers, pīwakawaka/New Zealand Fantails and weka. Various shorebirds forage on the beaches, and four species of shag and two species of penguin live and breed in Paterson Inlet.

## KEY FACTS

**Getting There**
Ulva Island is a short (6–10-minute) water-taxi ride from Golden Bay Wharf Rakiura/Stewart Island.

**Facilities**
Jetty with shelter, walking tracks, picnic shelter, cold running water and long drop toilets. Take your own food.

**Habitat**
Unmodified mature podocarp forest.

**Best Time to Visit**
Summer and early autumn. The winter months are often really tough on Rakiura/Stewart Island with howling winds and sleety rain. The tracks can be muddy. Summer is best but there is more bird activity in spring when there is mating and nesting.

Kelly Lynch

*Ulva Island, great for passerine spotting and shorebirds*

ewart Island Weka

## Walks

**lva's Guided Walks** (www.ulva.co.nz)
alf-day guided birding and botany walk, ffered all year but departing later in winter. ncludes boat transport to Ulva Island.

**uggedy Range** (www.ruggedyrange.com)
½ or 8½-hour guided walks with expert irding guides. Includes transport to Ulva land.

**elf-guided walks** Maps can be purchased for small fee at the information centre. Water axis can be ordered here too.

### Key Species

Stewart Island Kiwi, Black Shag, Pied Shag, Little Shag, Stewart Island Shag, Spotted Shag, Blue Shag, White-faced Heron, kōtuku/White Heron, Reef Heron, Cattle Egret, Australasian Bittern, Black Swan, Paradise Shelduck, Grey Duck, New Zealand Shoveler, Australasian Harrier Hawk, kārearea/New Zealand Falcon, Banded Rail, Stewart Island Weka, tōrea/ South Island Pied Oystercatcher, Variable Oystercatcher, Spur-winged Plover, Banded Dotterel, Southern New Zealand Dotterel, Eastern Bar-tailed Godwit, Ruddy Turnstone, Southern Skua, Southern Black-backed Gull, Red-billed Gull, Black-billed Gull, Black-fronted Tern, Caspian Tern, Antarctic Tern, White-fronted Tern, kererū/New Zealand Pigeon, South Island kākā, kākāriki/Red-crowned and Yellow-crowned Parakeets, Shining Cuckoo, Long-tailed Cuckoo, ruru/ Morepork, Sacred Kingfisher, South Island Rifleman, Welcome Swallow, New Zealand Pipit, pīwakawaka/New Zealand Fantail, Stewart Island Robin, Fernbird, pīpipi/Brown Creeper, mohua/Yellowhead, Grey Warbler, korimako/Bellbird, tūī, tauhou/Silvereye, tīeke/ South Island Saddleback. See www.ulva.co.nz/ bird-list.html. With thanks to Ron Tindal who compiled this list.

Oscar Thomas

## Mohua/Yellowhead

*Mohoua ochrocephala*

Yellowheads are endemic, small forest songsters about the size of a sparrow. As indicated by their name they have a vivid yellow head and underparts, and mottled brown wings and tail. They are vocal and flit about in groups, tweeting, chirping and trilling. They are flitters rather than fliers and tend not to fly long distances. Yellowheads, along with their close cousins, Whiteheads, and Brown Creepers are the key hosts of Long-tailed Cuckoos (see page 133), which lay their egg in their nests. Usually they have two clutches of eggs each season, so manage to get some of their own chicks fledged before they start caring for the cuckoo chick. They are susceptible to predation by rats and stoats, especially when nesting and roosting. They thrive in predator-free areas such as Ulva Island, and translocation to other islands and predator-free mainland sanctuaries has proved to be successful.

## Pīpipi/Brown Creeper

*Mohoua novaeseelandiae*

Pīpipi/Brown Creepers are small forest birc found only in the South Island in a variety forest types and scrubland areas. They hav grey heads and brown-mottled bodies, and are slightly bigger than Grey Warblers. The Brown Creeper is a squeaky tweeter and often flies in chattering flocks, providing a staccato addition to the dawn chorus. The birds have compact, cupped nests of twigs and moss lined with grass, feathers and wo and often have two clutches in spring and early summer. Both partners feed the chick and non-breeding birds will often help witl the task. Flocks of up to 60 form in autumn and summer and they will mob the much larger Long-tailed Cuckoo when it returns to the area in late spring. They are the main host for Long-tailed Cuckoos (see page 133).

### Accommodation

There is no accommodation on the island. See Rakiura/Stewart Island for accommodation options (page 207).

# Chatham Islands

*Chatham Island; pastureland, forest and a wild coastline*

Lying several hundred kilometres to the east of mainland New Zealand, the Chathams are a dozen or so volcanic islands and islets covering a total area of nearly 100,000ha. They emerged from the ocean about four million years ago, but in that time of oceanic isolation have played host to a fascinating process of island evolution. Today they are home to more than 50 endemic plant species, many specially adapted to the cool, windy climate, as well as a phenomenal cast of birds, crustaceans, freshwater fish, insects and other fauna.

Only the two largest islands, Chatham and Pitt (see page 215), are inhabited. When Moriori settled the Chathams in about AD 1500, they found islands largely covered with scrub, forest and swamp, and teeming with animal life, from invertebrates to sea mammals. The Europeans who settled from the 1790s onwards turned much of the land over to pasture; farm animals, along with the usual cargo of rodents, pigs, possums, cats and dogs, have done much harm. The small islands of Mangere and Rangatira (South East Island) are exceptions: nature reserves, cleared of livestock and pests, where rare native species are making a comeback. Both are strictly off-limits to visitors without special permits, but boat parties may observe the birdlife from offshore.

More than 50 native birds can be found in the Chathams. Along with millions-strong seabird colonies, there are 18 endemics, many of them endangered (some 14 species have become extinct since humans first arrived). Since much of the land on the Chathams is privately owned, you will need permission to visit certain sites, and even then there is no

## KEY FACTS

**Getting There**
There are flights to Tuuta Airport (20km out of Waitangi, the main settlement on Chatham Island) via Air Chathams from Auckland, Wellington and Christchurch on most days of the week. With no public transportation on the islands, your simplest option is to arrange everything – travel, transportation, accommodation and guided eco-tours – through a tour operator, who can place you with a local islander 'host'. If you do hire a car, note that most roads are unsealed. Local time is 45 minutes ahead of standard New Zealand time, and there is internet but no mobile phone coverage.

**Habitats**
Sea cliffs, rock stacks and islets, inland lagoons, forest, peat bog, pasture, scrub.

**Facilities**
Boat, car and bike hire, a visitor centre, a pub, a cafe or two, a grocery store and other small shops.

**Best Time to Visit**
Any time of year is good for birds, but winter squalls can hamper boat trips. The weather is variable throughout the year, so come prepared.

**Useful Websites**
www.newzealandbirds.co.nz/birding/chathamsbirds.html; www.doc.govt.nz/nature/habitats/offshore-islands/chatham-islands/chatham-islands-animals; www.taiko.org.nz; www.tourism.net.nz/new-zealand/about-new-zealand/regions/nz-outlying-islands/

guarantee of spotting your quarry. It can be hard, for instance, to catch a glimpse of the Chatham Island Robin, famously rescued from a low of five birds in 1980 to a population of more than 200 today.

## Tours

**Touring the Chathams** Standard tours of the main islands will take you to tourist sites such as the statue of Tommy Solomon, the restored flying boat, the beaches (for seal spotting, or combing for shells and sharks' teeth), and so on. A birding tour will typically focus on the bird-rich southwestern area of Chatham Island, a pelagic day for seabirds, and a visit to Pitt Island, as well as the waters off Mangere, Rangatira and The Pyramid. Either option may yield results: Chatham and Pitt Shags, for instance, can sometimes be seen just a couple of minutes' walk from the Solomon statue, and oystercatchers on Waitangi West beach. Mounting your own tour can be tricky. Boats may be privately hired, but at great cost. Also, you will need a special permit from the DOC, or permission from local landowners, to visit strongholds of the Chathams' rarer species,

### Key Species

Kororā/Little Penguin, Royal Albatross, Grey-headed Albatross/Mollymawk, Buller's Albatross, Shy Albatross/Mollymawk, Yellow-nosed Albatross, Northern Giant Petrel, Cape Pigeon, parea/Chatham Island Pigeon, Mottled Petrel, tāiko/Magenta Petrel, Soft-plumaged Petrel, Black-winged Petrel, Chatham Petrel, Broad-billed Prion, Fairy Prion, Fulmar Prion, tītī/ Sooty Shearwater/Muttonbird, Little Shearwater Grey-backed Storm Petrel, White-faced Storm Petrel, Southern Diving Petrel, Australasian Gannet, Black Shag, Chatham Shag, Pitt Shag, White-faced Heron, Reef Heron, Cattle Egret, Black Swan, Paradise Duck, Mallard, Grey Duck, New Zealand Shoveler, Swamp Harrier, Banded Rail, weka, Marsh Crake, Spotless Crake, pūkeko, tōrea/South Island Pied Oystercatcher, Chatham Island Oystercatcher, Pied Stilt, Spur-winged Plover, Banded Dotterel, New Zealand Shore Plover, Chatham Island Snipe, Eastern Bar-tailed Godwit, Turnstone, Red Knot, Arctic Skua, Southern Black-backed Gull, Red-billed Gull, White-fronted Tern, kākā, kākāriki/Red-crowned Parakeet, Long-tailed Cuckoo, Shining Cuckoo, pīwakawaka/New Zealand Fantail, Tomtit, Black Robin/Chatham Island Robin, Grey Warbler, Chatham Gerygone/Chatham Island Warbler, korimako/Bellbird, tūī, tauhou/Silvereye.

Ben Holyoake

*With its spectacular landscapes and changeable weather, remote Chatham Island is like no other place on earth*

ıereas tour operators can generally get you ›o such areas.

**ɣatotara Reserve, Tuku Farm Reserves, ›p Sanctuary, Sweetwater Predator-free ɣvenant** These areas all lie within the Tuanui ɱily's sheep farm on south-west Chatham, the easiest option is to book a 'Tuku Tour' ıth Liz and Bruce Tuanui. Tours run weekly, more often if demand is high, over summer. ıe Tuku Nature Reserve is a key breeding ɔund for the Chatham Island tāiko/Magenta ›trel. Under the aegis of the Tāiko Trust, e Sweetwater Covenant was fenced off in )06 to create a 2.4ha predator-proof 'fledging ırsery' for tāiko chicks translocated from eir breeding burrows in the Tuku Valley. now has 12 of these chicks breeding in ıere with other residents, including the ıatham Petrel and Diving Petrel; access Sweetwater is very limited, with a high e (which goes to the trust). The Tuku farm ›ea is the stronghold of the parea/Chatham Island Pigeon, which was nearly extinct in the late 1980s and is now abundant due to ongoing predator work and protection of bush. Depending on the season, tours take in Taiko Camp and the Gap Sanctuary, to which Chatham Island Albatross and, more recently, tītī chicks were transferred. Awatotara is open to unguided visitors (of reasonable fitness); it has two main tracks (2–3 hours and 3–4 hours) where you can see abundant endemics. See lbtuanui@farmside.co.nz.

**Te Whanga Lagoon** Due to the maritime climate, some 60 per cent of the Chathams' surface is dominated by peat soil, and the islands' wetlands are fabulously rich in plants, including 35 endemic species. Te Whanga, the 20,000ha lagoon at the heart of Chatham, is a great spot for waders and seabirds, including the Chatham Island Oystercatcher and Chatham and Pitt Shags.

## Chatham Island Albatross

*Thalassarche eremita*

With a population numbering around 5,300 pairs, the Chatham Island Albatross would seem to be relatively safe, but it is listed as vulnerable largely because it breeds on just one site: The Pyramid/Te Tara Koikoia, a huge conical rock stack to the south of Pitt Island. Severe storms, exacerbated by climate change, can shred the vegetation on The Pyramid, making nest sites precarious. To broaden the bird's chances of survival, in 2014 the Taiko Trust launched a project to establish a second breeding site at Point Gap on the south-west coast of Chatham Island. Volunteers set up a cluster of flowerpots to mimic the birds' mound nests, added plastic decoy birds to convey the sense of a busy colony, then began translocating chicks to this new site, and hand-fed them on fish and squid. Hundreds of chicks have been hand-reared on their flowerpots. Having fledged, they spend up to seven years at sea before returning to their colony to breed. At the time of writing, the trust hopes very soon to see the first hand-reared birds come back to Point Gap and raise their own chicks.

Oscar Thomas

## Chatham Island Snipe
*Coenocorypha pusilla*

The near extinction, then rescue, of the Chatham Island Robin is a well-known success story in New Zealand bird conservation, but another local bird came equally close to disappearing around 70 years ago. New Zealand's smallest snipe (indeed, the world's second smallest), the Chatham Island Snipe is a beautifully patterned bird, hard to spot in the dense undergrowth. Pairs nest in ground cover, often among tree roots, and share incubation of the two large eggs. When these hatch, the adults take scrupulous care of one chick each – you might call it 'solo parenting in a pair'. Today the snipe is reasonably numerous on Rangatira, Mangere, Little Mangere and Rabbit islands, as well as the Star Keys islets, but it was nearly eradicated from the Chathams by introduced rats, cats and weka. Fortunately it clung on in undisturbed Rangatira and the Star Keys, from where it was reintroduced to Mangere in 1970. Since then, it has extended its range and recovered to a total of perhaps 2,000 birds, but its extreme low numbers of the mid-twentieth century created a genetic bottleneck that renders the present population highly vulnerable to any introduced disease (as well as to cats and rats, should these predators ever succeed in recolonizing Rangatira and Mangere).

## Tāiko/Magenta Petrel
*Pterodroma magentae*

Numbering fewer than 200 birds, with barely more than a dozen breeding pairs, th Chatham Island tāiko, or Magenta Petrel, is one of the world's rarest seabirds. Indeed, fo a hundred years it was believed to be extinc until rediscovered in 1978. A medium-sized petrel with a dark head and pointed wings, it nests in ground burrows up to 5m long, solely in the dense forest of south-west Chatham Island. The tāiko's demise is almos entirely due to the arrival of humans. Cats, rats and pigs preyed on the bird at its nest, livestock trampled the burrows, and its breeding grounds were converted to pastur The Taiko Trust was formed in 2006 to brin landowners and conservationists together in efforts to rescue the species. It began by fencing off 2.4ha of land, now known as the Sweetwater Covenant, on Liz and Bruce Tuanui's farm in south-west Chatham, to create a predator-proof sanctuary. Chicks were translocated here to fledge, and in 2011 the first adults came back to breed at Sweetwater. The tāiko is still on the brink of extinction, but this vital lifeline may yet save it.

### Accommodation

There is everything from backpackers' facilities and motels, to homestays, self-catering guest houses, marae and hotels. Spaces are limited, so book before you travel.

# ›itt Island

*t and Mangere Islands from the air*

t 6,300ha, Pitt Island, which lies about 20km to the
outh-east of Chatham, is the second-largest island in the
rchipelago. Settled five centuries ago by the Moriori (who
amed it Rangiaotea/Rangihaute), Pitt was first visited
y British mariners in 1807, and by Māori (who named it
angiauria) a generation later. Today the island is home to
0–50 islanders. Most of the land is privately owned, with
bout 800ha tied up in scenic reserves.

'our best bet for visiting birding sites on Pitt is to book a
our, as most operators will have obtained special access from
andowners and the conservation estate. For instance, a tour
vith Flowerpot Bay Lodge (see Accommodation) can get you
nto reserves on the island, as well as boat trips off Mangere,
angatira (South East Island) and The Pyramid.

ome say that Pitt is better than Chatham for birdwatching;
ertainly it is more isolated, with fewer people to disturb the
ative wildlife. This, of course, means that accommodation is
imited, so you will need to plan ahead if you aim to include
'itt in a wider Chathams visit.

## KEY FACTS

**Getting There**
A six-seat Air Chathams Cessna makes regular flights on demand through the summer from Tuuta Airport (Chatham) to the Pitt airstrip. Boats also cross from Owenga (Chatham), taking roughly 50 minutes, but it is easier and generally cheaper to fly. Either way, book early. Bring all your essentials (particularly any medications) with you, too, as there are no shops on Pitt.

**Habitats**
Sea cliffs, rock stacks and islets, bush, pasture, scrub.

**Best Time to Visit**
Summer.

**Useful Websites**
www.newzealandbirds.co.nz/birding/chathamsbirds.html; www.doc.govt.nz/nature/habitats/offshore-islands/chatham-islands/chatham-islands-animals; www.tourism.net.nz/new-zealand/about-new-zealand/regions/nz-outlying-islands/nz-outlying-islands---chatham-islands/regional-information.html.

## Tracks and tours

**Flowerpot Bay tours** If you are staying at Flowerpot Bay Lodge, arranging a tour with them provides an easy way of seeing Pitt's birds. Three days is about right for most visitors, but they also do a day tour, which includes flights from/to Chatham and guided four-by-four tour of Pitt. Flowerpot arrange boat charter trips to the waters off Mangere and Rangatira (with a chance of seeing the New Zealand Shore Plover, Chatham Islands Parakeet and Pitt Shag), as well as The Pyramid, breeding site of the Chatham Island Albatross.

**Ellen Elizabeth Preece Conservation Covenant (Caravan Bush)** This 52ha predator-free DOC reserve in the north-east of Pitt contains two loop tracks, where you are likely to see pīwakawaka/New Zealand Fantail, Grey Warbler, kākāriki/Red-crowned Parakeet, Tomtit and tūī. If you visit during December–April, there is also a chance of seeing Chatham Petrels at their burrows. Access can be arranged with Flowerpot Bay Lodge.

### Key Species

Kororā/Little Penguin, Royal Albatross, Grey-headed Albatross/Mollymawk, Buller's Albatross, Shy Albatross/Mollymawk, Yellow-nosed Albatross, Northern Giant Petrel, Cape Pigeon, Mottled Petrel, tāiko/Magenta Petrel, Soft-plumaged Petrel, Black-winged Petrel, Chatham Petrel, Broad-billed Prion, Fairy Prion, Fulmar Prion, tītī/Sooty Shearwater, Little Shearwater, Grey-backed Storm Petrel, White-faced Storm Petrel, Southern Diving Petrel, Australasian Gannet, Black Shag, Chatham Shag, Pitt Shag, White-faced Heron, Reef Heron, Cattle Egret, Black Swan, Paradise Duck, Mallard, Grey Duck, New Zealand Shoveler, Swamp Harrier, California Quail, Banded Rail, weka, Marsh Crake, Spotless Crake, pūkeko, tōrea/South Island Pied Oystercatcher, Chatham Island Oystercatcher, poaka/Pied Stilt, Spur-winged Plover, Bander Dotterel, New Zealand Shore Plover, Chatham Island Snipe, Eastern Bar-tailed Godwit, Turnstone, Red Knot, Arctic Skua, Southern Black-backed Gull, Red-billed Gull, White-fronted Tern, kākā, kākāriki/Red-crowned Parakeet, Chatham Islands Parakeet, Long-tailed Cuckoo, Shining Cuckoo, pīwakawaka/New Zealand Fantail, Tomtit, Black Robin/Chatham Island Robin, Grey Warbler, Chatham Gerygone/Chatham Island Warbler, korimako/Bellbird, tūī, tauhou/Silvereye.

Brent Mallinson, Flowerpot Bay Lodge

*Flowerpot Bay Lodge*

## ack Robin

*etroica traversi*

the rarest bird in the world can be scued, then – given human determination d effort – no species need become extinct.' said the late Don Merton, a member of the Vildlife Service team that saved this little ack bird from oblivion in what is justifiably ne of the world's great conservation stories. y 1900, following the earlier introduction of ts and rats to the Chathams, the robin was xtinct on all islands except Little Mangere, 15ha slab lying off the tip of Mangere. y 1976, the Wildlife Service could find nly seven robins, and they moved these Mangere, where a further two died. By ow, salvation lay in just one breeding pair: female named Old Blue and her mate, Old ellow. At this point, Merton and his team ook the radical step of removing eggs from ld Blue to be fostered by Chatham Island omtits on Rangatira. Today the population is round 300, although it remains vulnerable o disease due to the low gene pool. So far, fforts to reintroduce robins to Pitt Island ave failed, but with further predator ontrol, success may yet come.

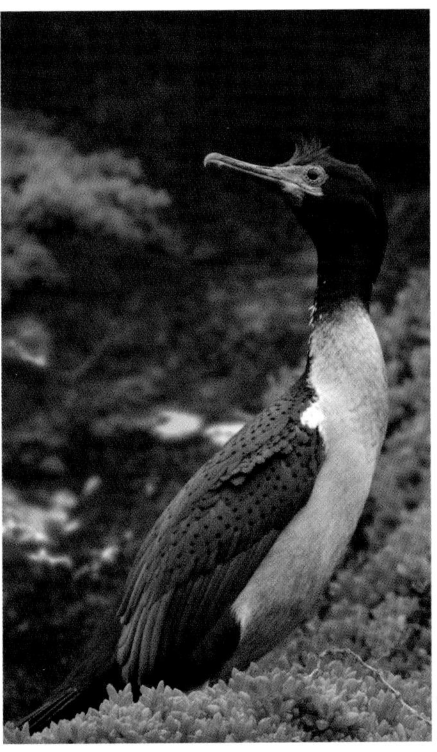

Oscar Thomas

## Pitt Shag

*Phalacrocorax featherstoni*

If you spot a slender grey shag with yellow feet in the Chathams, it is undoubtedly the endemic Pitt Shag. A relatively small bird, it is widely dispersed through the island group, but numbers are down to a few hundred pairs. The main factors in its decline are loss or disturbance of habitat, predation by introduced species, and fisheries bycatch; native gulls also prey on it. Pitt Shags roost in small groups on cliff faces and rock stacks, and tend to feed solo in coastal waters (rather than dispersing widely), so your best bet of spotting them is from a boat trip around Mangere or Rangatira.

## Accommodation

Flowerpot Bay Lodge, a stone's throw from Pitt's main wharf on the north coast, has room for 10 guests and is a key tour operator on Pitt. Other options include Rauceby and Hakepa homestays, both of which are based on farms and offer a range of activities.

## Further Information

Scofield, Paul & Stephenson, Brent. 2013. *Birds of New Zealand: a Photographic Guide*. Auckla University Press. Beautiful book.

Heather, Barrie & Robertson, Hugh. Illustrated by Derek Onley. Revised 2015. *The Field Guide To The Birds Of New Zealand*. Penguin Random House New Zealand, endorsed by the Ornithological Society of New Zealand.

**nzbirdsonline.org.nz** This searchable encyclopaedia of New Zealand birds is an incredible resource. It has detailed information about all New Zealand bird species. It has expert-writte texts, sound files of bird calls (lots of fun) and more than 8,700 photographs. And it is easy to navigate. The management of the website is covered by an agreement between Te Papa, Bir New Zealand (Ornithological Society of New Zealand) and DOC (Department of Conservatio

**www.birdingnz.co.nz** The New Zealand Birding Network is a marketing collective promotin birdwatching and bird-tour operators in New Zealand on a sustainable, ecological, commerci basis.

**www.nzbirds.com** Birders accommodation, birding sites, field guides, galleries, birding tours and even bird poems; this website has a lot to offer.

## Acknowledgements

Liz Light thanks her husband Sam Tracy who accompanied her to many of these sites; he set the alarm early in the morning, drove, and carried the picnic lunch and the water, while Liz took photographs and notes. Thank you, Sam, for your constant support and enthusiasm. Th would not have happened without you

Heartfelt thanks to the publishers in particular John Beaufoy and Rosemary Wilkinson. Thank you for your understanding and generosity in pushing this book across the line durin a traumatic time when it felt like completion was impossible. To Matt Turner who was roped in at the end and did a fine job, and to Krystyna Mayer for thorough editing and for persistin with the macrons on the many Māori words that need them.

And photography; so many people generously contributed images, too many to list them all here, but special thanks to Oscar Thomas. Oscar, at 18, is already a terrific bird photographer and ornithologist. He delivered beautiful images, of the requested size, and in a timely manne Thanks to John Kyngdon for his wonderful pelagic bird images; true seabirds are immensely difficult to find and photograph. Also, to Glenda Peake for her quirky bird imagery, Gert op den Dries for access to his collection, Kelly Lynch for her evocative photographs of Rakiura/ Stewart Island and Fiordland and to Paul Corliss for his Lake Ellesmere and Avon-Heathcote Estuary habitat images. And thanks to all the other bird enthusiasts who happily contributed those hard-to-get images.

# DEX



**te** The index includes species described in the text for each location. Page numbers for featured :cies with descriptions are in **bold**. Other illustrated species have *italic* page numbers. For :cies found at individual locations, refer to key species lists under specific locations within text.